TABLE 1.7

Approximate Physical Properties of Some Common Gases at Standard Atmospheric Pressure (BG Units)

Gas	Temperature (°F)	Density, ρ (slugs/ft^3)	Specific Weight, γ (lb/ft^3)	Dynamic Viscosity, μ (lb·s/ft^2)	Kinematic Viscosity, ν (ft^2/s)	Gas Constant,[a] R (ft·lb/slug·°R)	Specific Heat Ratio,[b] k
Air (standard)	59	2.38 E − 3	7.65 E − 2	3.74 E − 7	1.57 E − 4	1.716 E + 3	1.40
Carbon dioxide	68	3.55 E − 3	1.14 E − 1	3.07 E − 7	8.65 E − 5	1.130 E + 3	1.30
Helium	68	3.23 E − 4	1.04 E − 2	4.09 E − 7	1.27 E − 3	1.242 E + 4	1.66
Hydrogen	68	1.63 E − 4	5.25 E − 3	1.85 E − 7	1.13 E − 3	2.466 E + 4	1.41
Methane (natural gas)	68	1.29 E − 3	4.15 E − 2	2.29 E − 7	1.78 E − 4	3.099 E + 3	1.31
Nitrogen	68	2.26 E − 3	7.28 E − 2	3.68 E − 7	1.63 E − 4	1.775 E + 3	1.40
Oxygen	68	2.58 E − 3	8.31 E − 2	4.25 E − 7	1.65 E − 4	1.554 E + 3	1.40

[a]Values of the gas constant are independent of temperature.
[b]Values of the specific heat ratio depend only slightly on temperature.

TABLE 1.8

Approximate Physical Properties of Some Common Gases at Standard Atmospheric Pressure (SI Units)

Gas	Temperature (°C)	Density, ρ (kg/m^3)	Specific Weight, γ (N/m^3)	Dynamic Viscosity, μ (N·s/m^2)	Kinematic Viscosity, ν (m^2/s)	Gas Constant,[a] R (J/kg·K)	Specific Heat Ratio,[b] [illegible]
Air (standard)	15	1.23 E + 0	1.20 E + 1	1.79 E − 5	1.46 E − 5	2.869 E + 2	[illegible]
Carbon dioxide	20	1.83 E + 0	1.80 E + 1	1.47 E − 5	8.03 E − 6	1.889 E + 2	[illegible]
Helium	20	1.66 E − 1	1.63 E + 0	1.94 E − 5	1.15 E − 4	2.077 E + 3	[illegible]
Hydrogen	20	8.38 E − 2	8.22 E − 1	8.84 E − 6	1.05 E − 4	4.124 E + 3	[illegible]
Methane (natural gas)	20	6.67 E − 1	6.54 E + 0	1.10 E − 5	1.65 E − 5	5.183 E + 2	[illegible]
Nitrogen	20	1.16 E + 0	1.14 E + 1	1.76 E − 5	1.52 E − 5	2.968 E + 2	[illegible]
Oxygen	20	1.33 E + 0	1.30 E + 1	2.04 E − 5	1.53 E − 5	2.598 E + 2	[illegible]

[a]Values of the gas constant are independent of temperature.
[b]Values of the specific heat ratio depend only slightly on temperature.

Student Solutions Manual

Third Edition

Fundamentals of Fluid Mechanics

BRUCE R. MUNSON
DONALD F. YOUNG
Department of Aerospace Engineering and Engineering Mechanics
THEODORE H. OKIISHI
Department of Mechanical Engineering
Iowa State University
Ames, Iowa, USA

John Wiley & Sons, Inc.
New York Chichester Weinheim Brisbane Singapore Toronto

COVER PHOTOS Whirlpool: Francoise Sauze/Science Photo Library/Photo Researchers, Inc.
Computer Simulation of Vortex: Dr. Fred Espenak/Science Photo Library/Photo Researchers, Inc.

ACQUISITIONS EDITOR Charity Robey
MARKETING MANAGER Harper Mooy
PRODUCTION EDITOR Tony VenGraitis
COVER DESIGNER Madelyn Lesure

This book was printed and bound by Malloy Lithograph.
The cover was printed by Phoenix Color.

This book is printed on acid-free paper. ♾

The paper in this book was manufactured by a mill whose forest management programs include sustained yield harvesting of its timberlands. Sustained yield harvesting principles ensure that the number of trees cut each year does not exceed the amount of new growth.

ISBN 0-471-24011-7

Printed in the United States of America

10 9 8 7 6 5 4 3

PREFACE

This Student Solutions Manual has been developed as a supplement to the third edition of *Fundamentals of Fluid Mechanics* by Munson, Young, and Okiishi. In this third edition a new section has been added at the end of each chapter which contains a series of review problems. These problems are representative of the types of problems that students should be able to solve after completing the chapter, and this Solutions Manual contains the detailed solutions to these review problems. We believe that as students prepare for an examination, or feel the need for some additional work on a particular topic, it will be helpful to have available such a set of review problems with their corresponding solutions.

Each review problem is preceded by brief phrases which give an indication of the main topics to be used in solving the problem. Thus, the student can conveniently select those topics, and the corresponding review problems, of interest. This information is also presented in the table of contents. The solutions contained in this manual are worked in a logical, systematic way with sufficient detail so that they can be readily followed. Except where a greater accuracy is warranted, all intermediate calculations and answers are given to three significant figures. For convenience, some tables of properties and tables of unit conversion factors taken from *Fundamentals of Fluid Mechanics* are included on the inside of the front and back covers of this manual. Unless otherwise indicated in the problem statement, values of fluid properties used in the solutions are those given in the properties tables found on the inside of the front cover. Some occasional references to equations made in the solutions refer back to *Fundamentals of Fluid Mechanics*.

Although this Student Solutions Manual has been developed as a supplement to a particular text, the solutions are sufficiently detailed, using conventional notation and basic equations, so that the manual could also be a useful supplement to other standard fluid mechanics texts. Problem statements and figures are included with each solution.

The authors hope that this supplement to our text will be a useful tool to help the student gain a better understanding of basic fluid mechanics. We believe that practice through solving a variety of problems, with immediately available feedback by way of a Student Solution Manual, can be a valuable component in the spectrum of teaching tools needed in the study of fluid mechanics. Any suggestions and comments from you, the user, are certainly welcome and appreciated.

Bruce R. Munson
Donald F. Young
Theodore H. Okiishi

CONTENTS

1

Introduction

The breakup of a liquid jet into drops is a function of fluid properties such as density, viscosity, and surface tension. [Reprinted with permission from American Institute of Physics (Ref. 6) and the American Association for the Advancement of Science (Ref. 7).]

1.1R

1.1R (Dimensions) During a study of a certain flow system the following equation relating the pressures p_1 and p_2 at two points was developed:

$$p_2 = p_1 + \frac{f\ell V}{Dg}$$

In this equation V is a velocity, ℓ the distance between the two points, D a diameter, g the acceleration of gravity, and f a dimensionless coefficient. Is the equation dimensionally consistent?

(ANS: No)

$$p_2 = p_1 + \frac{f\ell V}{Dg}$$

$$[FL^{-2}] \doteq [FL^{-2}] + \left[\frac{(F^0L^0T^0)(L)(LT^{-1})}{(L)(LT^{-2})}\right]$$

$$[FL^{-2}] \doteq [FL^{-2}] + [T]$$

Since each term in the equation does not have the same dimensions, the equation is not dimensionally consistent. No.

1.2R

1.2R (Dimensions) If V is a velocity, ℓ a length, $\mathcal{W}$ a weight, and μ a fluid property having dimensions of $FL^{-2}T$, determine the dimensions of: **(a)** $V\ell\mathcal{W}/\mu$, **(b)** $\mathcal{W}\mu\ell$, **(c)** $V\mu/\ell$, and **(d)** $V\ell^2\mu/\mathcal{W}$.

(ANS: L^4T^{-2}; $F^2L^{-1}T$; FL^{-2}; L)

(a) $$\frac{V\ell\mathcal{W}}{\mu} \doteq \frac{(LT^{-1})(L)(F)}{(FL^{-2}T)} \doteq L^4T^{-2}$$

(b) $$\mathcal{W}\mu\ell \doteq (F)(FL^{-2}T)(L) \doteq F^2L^{-1}T$$

(c) $$\frac{V\mu}{\ell} \doteq \frac{(LT^{-1})(FL^{-2}T)}{(L)} \doteq FL^{-2}$$

(d) $$\frac{V\ell^2\mu}{\mathcal{W}} \doteq \frac{(LT^{-1})(L^2)(FL^{-2}T)}{(F)} \doteq L$$

1.3R

1.3R (Units) Make use of Table 1.4 to express the following quantities in BG units: **(a)** 465 W, **(b)** 92.1 J, **(c)** 536 N/m^2, **(d)** 85.9 mm^3, **(e)** 386 kg/m^2.

(ANS: 3.43×10^2 ft·lb/s; 67.9 ft·lb; 11.2 lb/ft^2; 3.03×10^{-6} ft^3; 2.46 $slugs/ft^2$)

$$(a)\quad 465\ W = (465\ W)\left(7.376 \times 10^{-1}\ \frac{\frac{ft \cdot lb}{s}}{W}\right) = \underline{\underline{3.43 \times 10^2\ \frac{ft \cdot lb}{s}}}$$

$$(b)\quad 92.1\ J = (92.1\ J)\left(7.376 \times 10^{-1}\ \frac{ft \cdot lb}{J}\right) = \underline{\underline{67.9\ ft \cdot lb}}$$

$$(c)\quad 536\ \frac{N}{m^2} = \left(536\ \frac{N}{m^2}\right)\left(2.089 \times 10^{-2}\ \frac{\frac{lb}{ft^2}}{\frac{N}{m^2}}\right) = \underline{\underline{11.2\ \frac{lb}{ft^2}}}$$

$$(d)\quad 85.9\ mm^3 = \left(85.9 \times 10^{-9}\ m^3\right)\left[(3.281)^3\ \frac{ft^3}{m^3}\right] = \underline{\underline{3.03 \times 10^{-6}\ ft^3}}$$

$$(e)\quad 386\ \frac{kg}{m^2} = \left(386\ \frac{kg}{m^2}\right)\left(6.852 \times 10^{-2}\ \frac{slugs}{kg}\right)\left(\frac{1}{1.076 \times 10\ \frac{ft^2}{m^2}}\right)$$

$$= \underline{\underline{2.46\ \frac{slugs}{ft^2}}}$$

1.4R

1.4R (Units) A person weighs 165 lb at the earth's surface. Determine the person's mass in slugs, kilograms, and pounds mass.

(ANS: 5.12 slugs; 74.8 kg; 165 lbm)

$$mass = \frac{Weight}{g}$$

In slugs,

$$mass = \frac{165\ lb}{32.2\ ft/s^2} = \underline{\underline{5.12\ slugs}}$$

In kg,

$$mass = \frac{(165\ lb)\left(4.448\ \frac{N}{lb}\right)}{9.81\ m/s^2} = \underline{\underline{74.8\ kg}}$$

In lbm,

$$mass = \underline{\underline{165\ lbm}}$$

1.5R

1.5R (Specific gravity) Make use of Fig. 1.1 to determine the specific gravity of water at 22 and 89 °C. What is the specific volume of water at these two temperatures?

(ANS: 0.998; 0.966; $1.002 \times 10^{-3} m^3/kg$; $1.035 \times 10^{-3} m^3/kg$)

From Fig. 1.1:

at 22°C $\quad \rho_{H_2O} = 998 \frac{kg}{m^3}$

at 89°C $\quad \rho_{H_2O} = 966 \frac{kg}{m^3}$

Thus,

$$SG = \frac{\rho}{\rho_{H_2O} @ 4°C} = \frac{998 \frac{kg}{m^3}}{1000 \frac{kg}{m^3}} = \underline{\underline{0.998 @ 22°C}}$$

and

$$SG = \frac{966 \frac{kg}{m^3}}{1000 \frac{kg}{m^3}} = \underline{\underline{0.966 @ 89°C}}$$

Since specific volume $v = \frac{1}{\rho}$

$$v = \frac{1}{998 \frac{kg}{m^3}} = \underline{\underline{1.002 \times 10^{-3} \frac{m^3}{kg} \quad @ \quad 22°C}}$$

and

$$v = \frac{1}{966 \frac{kg}{m^3}} = \underline{\underline{1.035 \times 10^{-3} \frac{m^3}{kg} \quad @ \quad 89°C}}$$

1.6R

1.6R (Specific weight) A 1-ft-diameter cylindrical tank that is 5 ft long weighs 125 lb and is filled with a liquid having a specific weight of 69.6 lb/ft^3. Determine the vertical force required to give the tank an upward acceleration of 9 ft/s^2.
(ANS: 509 lb up)

weight of tank $= W_t = 125\text{ lb}$

weight of liquid $= W_\ell = \gamma_\ell \times \forall$

$$W_\ell = \left(69.6\ \frac{\text{lb}}{\text{ft}^3}\right)\left[\frac{\pi}{4}(1\text{ ft}^2)(5\text{ ft})\right]$$

$$= 273\text{ lb}$$

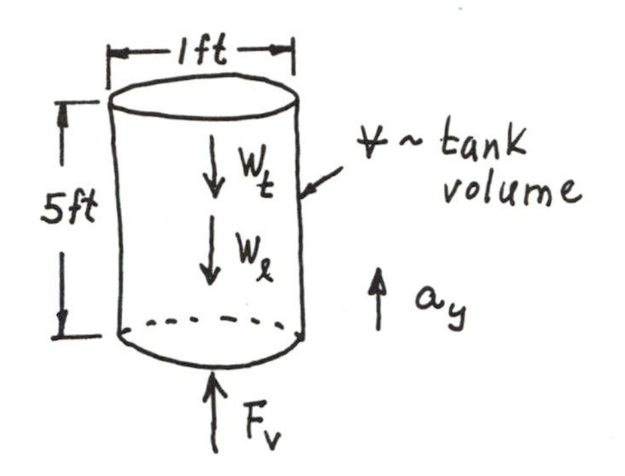

$(\uparrow +)$ $\sum F_y = m\,a_y$

$$F_v - W_t - W_\ell = m\,a_y$$

$$F_v - 125\text{ lb} - 273\text{ lb} = \left(\frac{125\text{ lb} + 273\text{ lb}}{32.2\ \frac{\text{ft}}{\text{s}^2}}\right) \times 9\ \frac{\text{ft}}{\text{s}^2}$$

$$F_v = 509\text{ lb} \uparrow$$

1.7R

1.7R (Ideal gas law) Calculate the density and specific weight of air at a gage pressure of 100 psi and a temperature of 100 °F. Assume standard atmospheric pressure.
(ANS: 1.72×10^{-2} slugs/ft^3; 0.554 lb/ft^3)

From the ideal gas law:

$$\rho = \frac{p}{RT} = \frac{\left(100\ \frac{\text{lb}}{\text{in.}^2} + 14.7\ \frac{\text{lb}}{\text{in.}^2}\right)\left(\frac{144\text{ in.}^2}{\text{ft}^2}\right)}{\left(1.716\times10^3\ \frac{\text{ft}\cdot\text{lb}}{\text{slug}\cdot{}^\circ\text{R}}\right)\left[(100\,^\circ\text{F} + 460)\,^\circ\text{R}\right]}$$

$$= 1.72 \times 10^{-2}\ \frac{\text{slugs}}{\text{ft}^3}$$

$$\gamma = \rho g = \left(1.72\times10^{-2}\ \frac{\text{slugs}}{\text{ft}^3}\right)\left(32.2\ \frac{\text{ft}}{\text{s}^2}\right) = 0.554\ \frac{\text{lb}}{\text{ft}^3}$$

1.8R

1.8R (Ideal gas law) A large dirigible having a volume of 90,000 m^3 contains helium under standard atmospheric conditions [pressure = 101 kPa (abs) and temperature = 15 °C]. Determine the density and total weight of the helium.

(ANS: 0.169 kg/m^3; 1.49 × 10^5N)

volume = 90,000 m^3

From the ideal gas law,

$$\rho = \frac{p}{RT} = \frac{101 \times 10^3 \frac{N}{m^2}}{\left(2077 \frac{J}{kg \cdot K}\right)\left[(15°C + 273)K\right]} = \underline{\underline{0.169 \frac{kg}{m^3}}}$$

$$\text{weight} = \rho g \times \text{volume} = \left(0.169 \frac{kg}{m^3}\right)\left(9.81 \frac{m}{s^2}\right)\left(9 \times 10^4 \, m^3\right)$$

$$= \underline{\underline{1.49 \times 10^5 \, N}}$$

1.9R

1.9R (Viscosity) A Newtonian fluid having a specific gravity of 0.92 and a kinematic viscosity of 4 × 10^{-4} m^2/s flows past a fixed surface. The velocity profile near the surface is shown in Fig. P1.9R. Determine the magnitude and direction of the shearing stress developed on the plate. Express your answer in terms of U and δ, with U and δ expressed in units of meters per second and meters, respectively.

(ANS: 0.578 U/ δ N/m^2 acting to right on plate)

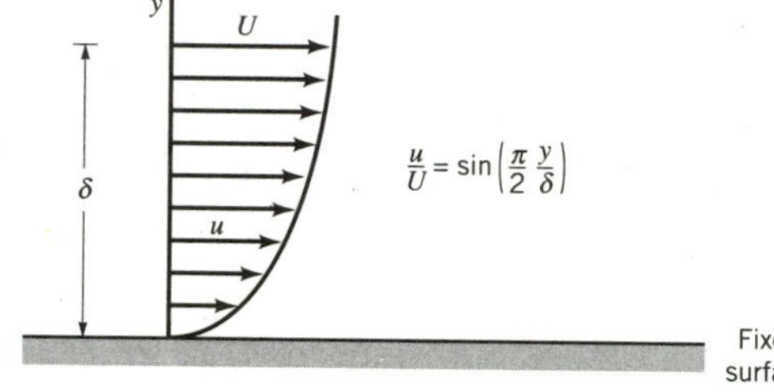

■ FIGURE P1.9R

$$\tau_{\substack{surface \\ (y=0)}} = \mu \left(\frac{du}{dy}\right)_{y=0} \quad \text{where } \mu = \nu \rho$$

$$\frac{du}{dy} = \frac{\pi}{2}\frac{U}{\delta} \cos\left(\frac{\pi}{2}\frac{y}{\delta}\right)$$

At $y = 0$, $\frac{du}{dy} = \frac{\pi}{2}\frac{U}{\delta}(1)$

Since, $\mu = \nu\rho$ where $\rho = SG \, \rho_{H_2O} = 0.92\left(1000 \frac{kg}{m^3}\right)$

$$\tau_{surface} = \nu\rho\left(\frac{\pi}{2}\frac{U}{\delta}\right)$$

$$= \left(4 \times 10^{-4} \frac{m^2}{s}\right)\left(0.92 \times 10^3 \frac{kg}{m^3}\right)\left(\frac{\pi}{2}\right)\frac{U}{\delta}$$

$$= \underline{\underline{0.578 \frac{U}{\delta} \; N/m^2 \text{ acting to right on plate}}}$$

1.10R

1.10R (Viscosity) A large movable plate is located between two large fixed plates as shown in Fig. P1.10R. Two Newtonian fluids having the viscosities indicated are contained between the plates. Determine the magnitude and direction of the shearing stresses that act on the fixed walls when the moving plate has a velocity of 4 m/s as shown. Assume that the velocity distribution between the plates is linear.

(ANS: 13.3 N/m^2 in direction of moving plate)

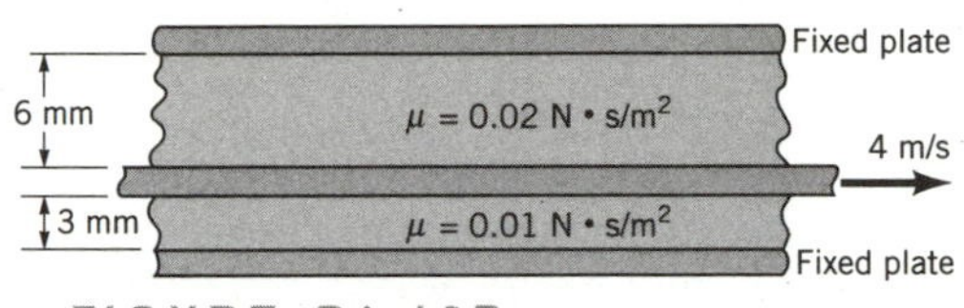

■ FIGURE P1.10R

$$\tau = \mu \frac{du}{dy} = \mu \frac{U}{b} \text{ so that}$$

$$\tau_1 = \mu_1 \frac{U}{b_1} = \left(0.02 \frac{N \cdot s}{m^2}\right)\left(\frac{4 \frac{m}{s}}{0.006 m}\right)$$

$$= \underline{\underline{13.3 \frac{N}{m^2}}}$$

$$\tau_2 = \mu_2 \frac{U}{b_2} = \left(0.01 \frac{N \cdot s}{m^2}\right)\left(\frac{4 \frac{m}{s}}{0.003 m}\right)$$

$$= \underline{\underline{13.3 \frac{N}{m^2}}}$$

Stresses act on fixed walls in direction of moving plate.

1.11R

1.11R (Viscosity) Determine the torque required to rotate a 50-mm-diameter vertical cylinder at a constant angular velocity of 30 rad/s inside a fixed outer cylinder that has a diameter of 50.2 mm. The gap between the cylinders is filled with SAE 10 oil at 20 °C. The length of the inner cylinder is 200 mm. Neglect bottom effects and assume the velocity distribution in the gap is linear. If the temperature of the oil increases to 80 °C, what will be the percentage change in the torque?

(ANS: 0.589 N m; 92.0 percent)

Torque, $d\mathcal{T}$, due to shearing stress on inner cylinder is equal to

$$d\mathcal{T} = R_i \tau \, dA$$

where $dA = (R_i \, d\theta)\ell$. Thus,

$$d\mathcal{T} = R_i^2 \ell \tau \, d\theta$$

and torque required to rotate inner cylinder is

$$\mathcal{T} = R_i^2 \ell \tau \int_0^{2\pi} d\theta$$

$$= 2\pi R_i^2 \ell \tau$$

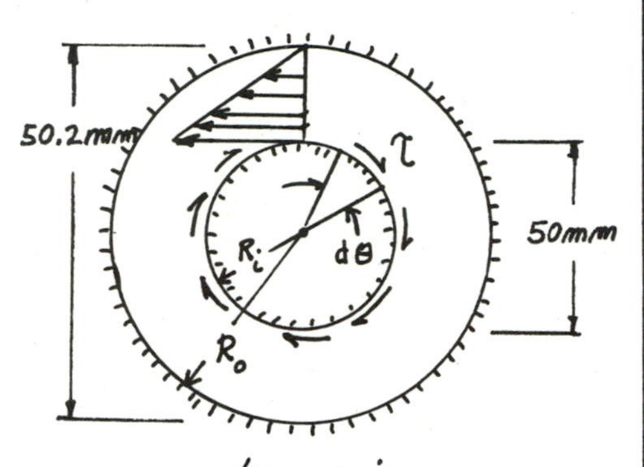

top view

(ℓ ~ cylinder length) = 200mm

For a linear velocity distribution in the gap,

$$\tau = \mu \frac{R_i \omega}{R_o - R_i}$$

so that

$$\mathcal{T} = \frac{2\pi R_i^3 \ell \mu \omega}{R_o - R_i}$$

From Fig. B.1 in Appendix B:

(for SAE 10 oil at 20°C) $\mu = 1.0 \times 10^{-1} \frac{N \cdot s}{m^2}$

(for SAE 10 oil at 80°C) $\mu = 8.0 \times 10^{-3} \frac{N \cdot s}{m^2}$

(contined)

1.11R continued

Thus, at 20°C

$$\mathcal{T} = \frac{2\pi\left(\frac{0.050}{2}\,m\right)^3(0.2\,m)\left(1.0\times10^{-1}\,\frac{N\cdot s}{m^2}\right)\left(30\,\frac{rad}{s}\right)}{\left(\frac{0.0502\,m - 0.0500\,m}{2}\right)}$$

$$= \underline{\underline{0.589\ N\cdot m}}$$

At 80°C,

$$\mathcal{T} = (0.589\ N\cdot m)\,\frac{\mu_{80^\circ}}{\mu_{20^\circ}}$$

so that % reduction in torque is equal to:

$$\%\text{ reduction in } \mathcal{T} = \left[\frac{0.589 - 0.589\,\frac{\mu_{80^\circ}}{\mu_{20^\circ}}}{0.589}\right]\times 100$$

$$= \left[1 - \frac{8.0\times10^{-3}}{1.0\times10^{-1}}\right]\times 100 = \underline{\underline{92.0\%}}$$

1.12R

1.12R (Bulk modulus) Estimate the increase in pressure (in psi) required to decrease a unit volume of mercury by 0.1%. (ANS: 4.14×10^3 psi)

$$E_v = -\frac{dp}{d\mathcal{V}/\mathcal{V}}$$, where from Table 1.5, $E_v = 4.14\times10^6\ \frac{lb}{in.^2}$

Thus,

$$\Delta p \approx -\frac{E_v\,\Delta\mathcal{V}}{\mathcal{V}} = -\left(4.14\times10^6\ \frac{lb}{in.^2}\right)(-0.001)$$

$$\Delta p \approx \underline{\underline{4.14\times10^3\ psi}}$$

1.13R

1.13R (Bulk modulus) What is the isothermal bulk modulus of nitrogen at a temperature of 90 °F and an absolute pressure of 5600 lb/ft^2?
(ANS: 5600 lb/ft^2)

For isothermal bulk modulus,

$$E_v = p \qquad \text{(Eq. 1.16)}$$

so that

$$E_v = 5600 \ \frac{\text{lb}}{\text{ft}^2}$$

1.14R

1.14R (Speed of sound) Compare the speed of sound in mercury and oxygen at 20 °C.
(ANS: $c_{Hg}/c_{O2} = 4.45$)

For mercury,

$$c = \sqrt{\frac{E_v}{\rho}} \qquad \text{(Eq. 1.19)}$$

so that

$$c = \sqrt{\frac{2.85 \times 10^{10} \ \frac{N}{m^2}}{1.36 \times 10^4 \ \frac{kg}{m^3}}} = 1.45 \times 10^3 \ \frac{m}{s}$$

For oxygen,

$$c = \sqrt{kRT} \qquad \text{(Eq. 1.20)}$$

so that

$$c = \sqrt{(1.40)\left(259.8 \ \frac{J}{kg \cdot K}\right)\left[(20°C + 273)K\right]} = 326 \ \frac{m}{s}$$

Thus,

$$\frac{c\,(\text{mercury})}{c\,(\text{oxygen})} = \frac{1.45 \times 10^3 \ \frac{m}{s}}{326 \ \frac{m}{s}} = 4.45$$

(See Tables 1.6 and 1.8 for values of E_v, k, and R.)

1.15R

1.15R (Vapor pressure) At a certain altitude it was found that water boils at 90 °C. What is the atmospheric pressure at this altitude?

(ANS: 70.1 kPa (abs))

The vapor pressure of water at 90°C is 7.01×10^4 Pa (abs) (from Table B.2 in Appendix B). Thus, if water boils at this temperature, the atmospheric pressure must be equal to

$$p_{atm} = 7.01 \times 10^4 \text{ Pa} = \underline{\underline{70.1 \text{ kPa (abs)}}}$$

2
Fluid Statics

An image of hurricane Allen viewed via satellite: Although there is considerable motion and structure to a hurricane, the pressure variation in the vertical direction is approximated by the pressure-depth relationship for a static fluid. (Visible and infrared image pair from a NOAA satellite using a technique developed at NASA/GSPC.) (Photograph courtesy of A. F. Hasler [Ref. 7].)

2.1R

2.1R (Pressure head) Compare the column heights of water, carbon tetrachloride, and mercury corresponding to a pressure of 50 kPa. Express your answer in meters.

(ANS: 5.10 m; 3.21 m; 0.376 m)

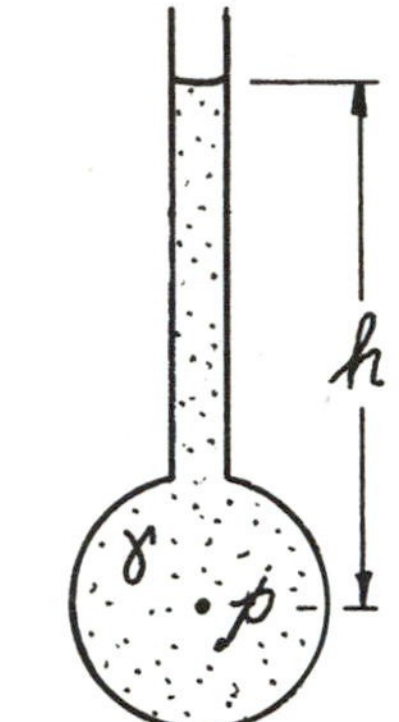

$$p = \gamma h$$

For water:
$$h = \frac{50 \times 10^3 \frac{N}{m^2}}{9.80 \times 10^3 \frac{N}{m^3}} = \underline{\underline{5.10\ m}}$$

For carbon tetrachloride:
$$h = \frac{50 \times 10^3 \frac{N}{m^2}}{15.6 \times 10^3 \frac{N}{m^3}} = \underline{\underline{3.21\ m}}$$

For mercury:
$$h = \frac{50 \times 10^3 \frac{N}{m^2}}{133 \times 10^3 \frac{N}{m^3}} = \underline{\underline{0.376\ m}}$$

2.2R

2.2R (Pressure-depth relationship) A closed tank is partially filled with glycerin. If the air pressure in the tank is 6 lb/in.^{2} and the depth of glycerin is 10 ft, what is the pressure in lb/ft^2 at the bottom of the tank?

(ANS: 1650 lb/ft^2)

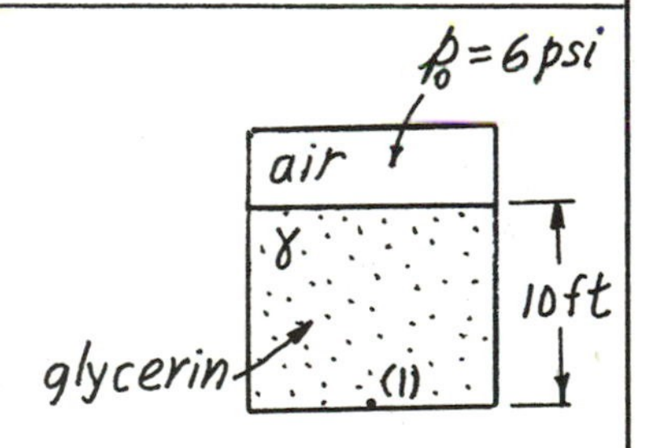

$$p_1 = \gamma h + p_o = \left(78.6 \frac{lb}{ft^3}\right)(10\ ft) + \left(6 \frac{lb}{in.^2}\right)\left(\frac{144\ in.^2}{ft^2}\right)$$

$$= \underline{\underline{1650 \frac{lb}{ft^2}}}$$

2.3R

2.3R (Gage-absolute pressure) On the inlet side of a pump a Bourdon pressure gage reads 600 lb/ft^2 vacuum. What is the corresponding absolute pressure if the local atmospheric pressure is 14.7 psia?

(ANS: 10.5 psia)

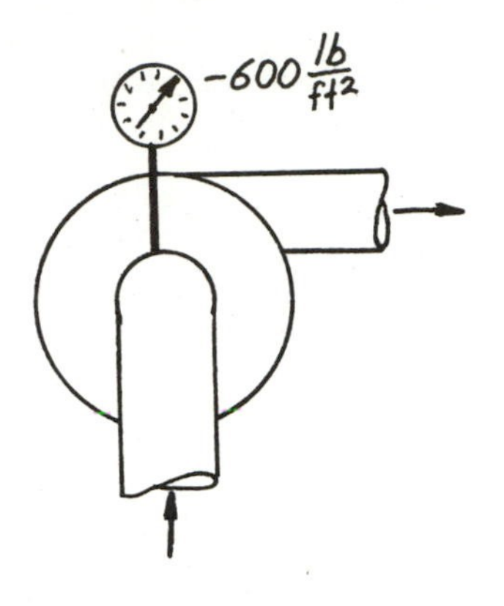

$$p(abs) = p(gage) + p(atm)$$

$$= -600 \frac{lb}{ft^2}\left(\frac{1\ ft^2}{144\ in.^2}\right) + 14.7\ psia$$

$$= \underline{\underline{10.5\ psia}}$$

2.4R

2.4R (Manometer) A tank is constructed of a series of cylinders having diameters of 0.30, 0.25, and 0.15 m as shown in Fig. P2.4R. The tank contains oil, water, and glycerin and a mercury manometer is attached to the bottom as illustrated. Calculate the manometer reading, h.

(ANS: 0.0327 m)

■ FIGURE P2.4R

$$p_1 + \gamma_{oil}(0.1m) + \gamma_{H_2O}(0.1m) + \gamma_{gly}(0.2m) - \gamma_{Hg}\,h = p_2$$

Thus, with $p_1 = p_2 = 0$,

$$h = \frac{\left(8.95\ \frac{kN}{m^3}\right)(0.1m) + \left(9.80\ \frac{kN}{m^3}\right)(0.1m) + \left(12.4\ \frac{kN}{m^3}\right)(0.2m)}{133\ \frac{kN}{m^3}}$$

$$= \underline{\underline{0.0327\ m}}$$

2.5R

2.5R (Manometer) A mercury manometer is used to measure the pressure difference in the two pipelines of Fig. P2.5R. Fuel oil (specific weight = 53.0 lb/ft³) is flowing in *A* and SAE 30 lube oil (specific weight = 57.0 lb/ft³) is flowing in *B*. An air pocket has become entrapped in the lube oil as indicated. Determine the pressure in pipe *B* if the pressure in *A* is 15.3 psi.

(ANS: 18.2 psi)

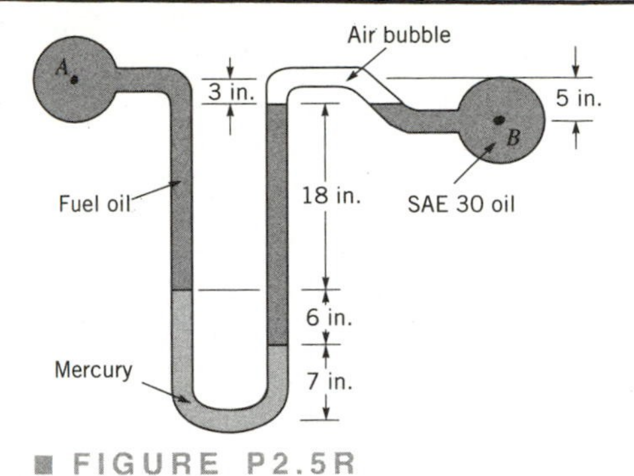

■ FIGURE P2.5R

$$p_A + \gamma_{fuel\ oil}\left(\frac{3+18}{12}\,ft\right) + \gamma_{Hg}\left(\frac{6}{12}\,ft\right) - \gamma_{SAE\,30}\left(\frac{6+18}{12}\,ft\right) + \gamma_{SAE\,30}\left(\frac{2}{12}\,ft\right) = p_B$$

Thus,

$$p_B = \left(15.3\,\frac{lb}{in.^2}\right)\left(144\,\frac{in^2}{ft^2}\right) + \left(53.0\,\frac{lb}{ft^3}\right)\left(\frac{21}{12}\,ft\right) + \left(847\,\frac{lb}{ft^3}\right)\left(\frac{6}{12}\,ft\right) - \left(57.0\,\frac{lb}{ft^3}\right)\left(\frac{22}{12}\,ft\right)$$

$$= 2615\,\frac{lb}{ft^2} = \left(2615\,\frac{lb}{ft^2}\right)\left(\frac{1\,ft^2}{144\,in.^2}\right) = \underline{\underline{18.2\ psi}}$$

2.6R

2.6R (Manometer) Determine the angle θ of the inclined tube shown in Fig. P2.6R if the pressure at *A* is 1 psi greater than that at *B*.

(ANS: 19.3 deg)

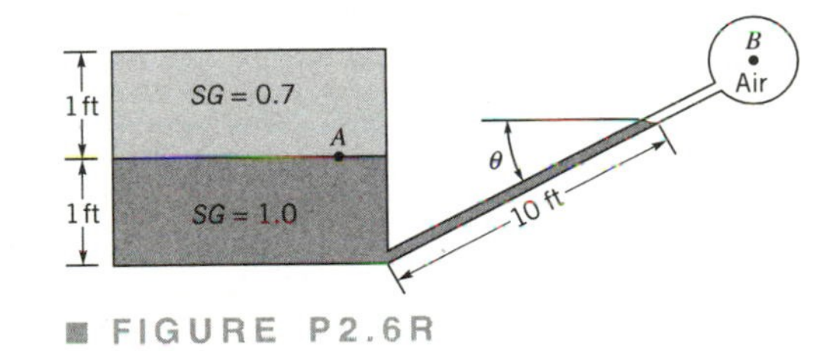

■ FIGURE P2.6R

$$p_A + (1.0)\left(62.4\,\frac{lb}{ft^3}\right) - (1.0)\left(62.4\,\frac{lb}{ft^3}\right)(10\,ft)\sin\theta = p_B$$

Thus,

$$p_A - p_B = (1.0)\left(62.4\,\frac{lb}{ft^3}\right)\left[(10\,ft)\sin\theta - 1\right]$$

Since $p_A - p_B = 1\ psi$

$$(10\,ft)\sin\theta - 1 = \frac{\left(1\,\frac{lb}{in.^2}\right)\left(144\,\frac{in.^2}{ft^2}\right)}{(1.0)\left(62.4\,\frac{lb}{ft^3}\right)}$$

so that

$$\sin\theta = 0.331 \quad \text{or} \quad \underline{\underline{\theta = 19.3^\circ}}$$

2.7R

2.7R (Force on plane surface) A swimming pool is 18 m long and 7 m wide. Determine the magnitude and location of the resultant force of the water on the vertical end of the pool where the depth is 2.5 m.

(ANS: 214 kN on centerline, 1.67 m below surface)

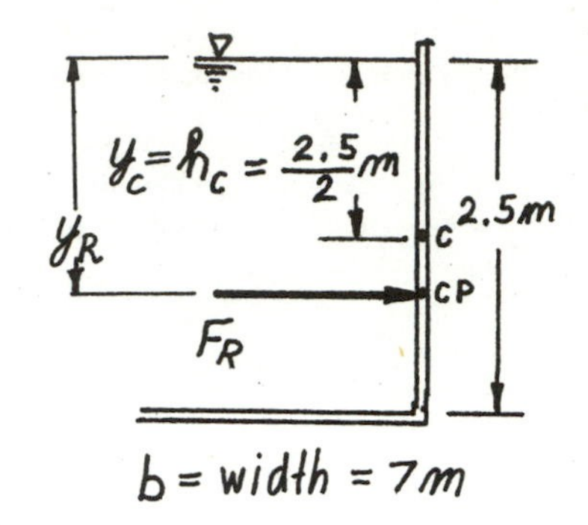

$$F_R = \gamma h_c A = \left(9.80\ \frac{kN}{m^3}\right)\left(\frac{2.5\,m}{2}\right)(7m \times 2.5m) = \underline{\underline{214\ kN}}$$

$$y_R = \frac{I_{xc}}{y_c A} + y_c \qquad \text{where} \quad I_{xc} = \frac{1}{12}(7m)(2.5m)^3$$

Thus,

$$y_R = \frac{\frac{1}{12}(7m)(2.5m)^3}{\left(\frac{2.5m}{2}\right)(7m \times 2.5m)} + \frac{2.5m}{2} = \underline{\underline{1.67\ m}}$$

The force of 214 kN acts 1.67 m below surface along vertical centerline of end.

2.8R

2.8R (Force on plane surface) The vertical cross section of a 7-m-long closed storage tank is shown in Fig. P2.8R. The tank contains ethyl alcohol and the air pressure is 40 kPa. Determine the magnitude of the resultant fluid force acting on one end of the tank.

(ANS: 847 kN)

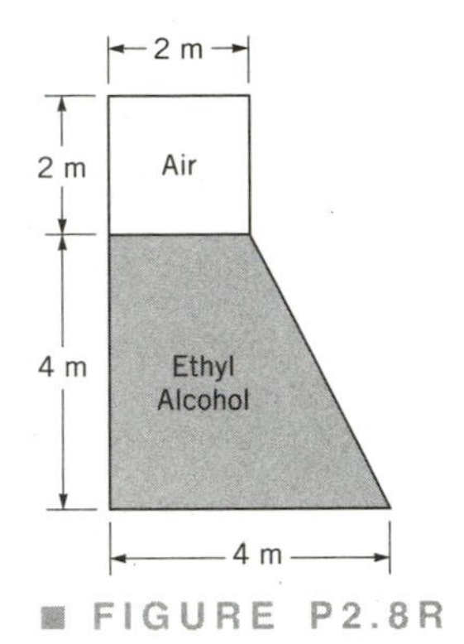

■ FIGURE P2.8R

Break area into three parts as shown in figure.

For area 1:

$$F_{R1} = p_{air} A_1 = \left(40 \frac{kN}{m^2}\right)(2m \times 2m) = 160 \text{ kN}$$

For area 2: (From Table 1.6 $\gamma_{\text{ethyl alcohol}} = 7.74 \frac{kN}{m^3}$)

$$F_{R2} = p_{air} A_2 + \gamma h_{c2} A_2$$

$$= \left(40 \frac{kN}{m^2}\right)(2m \times 4m) + \left(7.74 \frac{kN}{m^3}\right)\left(\frac{4m}{2}\right)(2m \times 4m)$$

$$= 444 \text{ kN}$$

For area 3:

$$F_{R3} = p_{air} A_3 + \gamma h_{c3} A_3$$

$$= \left(40 \frac{kN}{m^2}\right)\left(\frac{1}{2}\right)(2m \times 4m) + \left(7.74 \frac{kN}{m^3}\right)\left(\frac{2}{3}\right)(4m)\left(\frac{1}{2}\right)(2m \times 4m)$$

$$= 243 \text{ kN}$$

Thus,

$$F_R = F_{R1} + F_{R2} + F_{R3}$$

$$= 160 \text{ kN} + 444 \text{ kN} + 243 \text{ kN} = \underline{\underline{847 \text{ kN}}}$$

2.9R

2.9R (Center of pressure) A 3-ft-diameter circular plate is located in the vertical side of an open tank containing gasoline. The resultant force that the gasoline exerts on the plate acts 3.1 in. below the centroid of the plate. What is the depth of the liquid above the centroid?

(ANS: 2.18 ft)

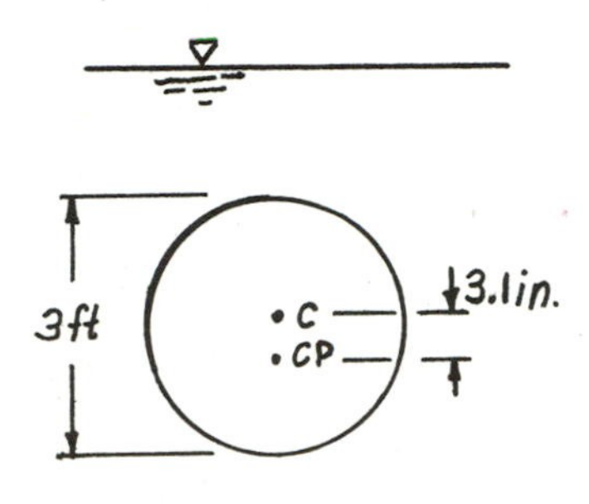

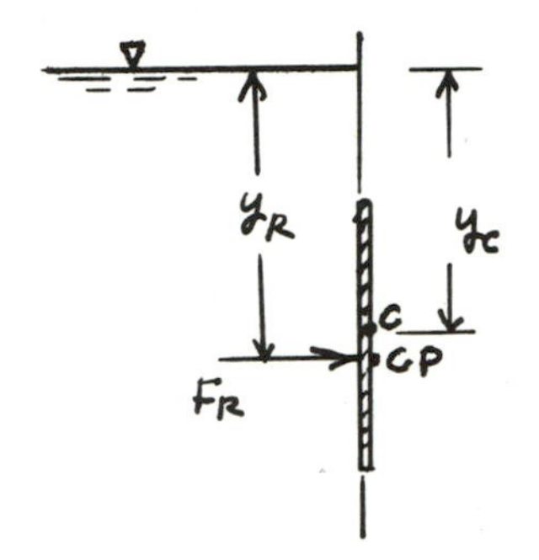

$$y_R = \frac{I_{xc}}{y_c A} + y_c \quad \text{where} \quad I_{xc} = \frac{\pi}{4}\left(\frac{3}{2}\text{ft}\right)^4$$

Thus,

$$y_R - y_c = \frac{3.1}{12}\text{ ft} = \frac{\frac{\pi}{4}\left(\frac{3}{2}\text{ ft}\right)^4}{y_c \frac{\pi}{4}(3\text{ft})^2}$$

so that

$$y_c = \underline{\underline{2.18 \text{ ft}}}$$

2.10R

2.10R (Force on plane surface) A gate having the triangular shape shown in Fig. P2.10R is located in the vertical side of an open tank. The gate is hinged about the horizontal axis AB. The force of the water on the gate creates a moment with respect to the axis AB. Determine the magnitude of this moment.

(ANS: 3890 kN·m)

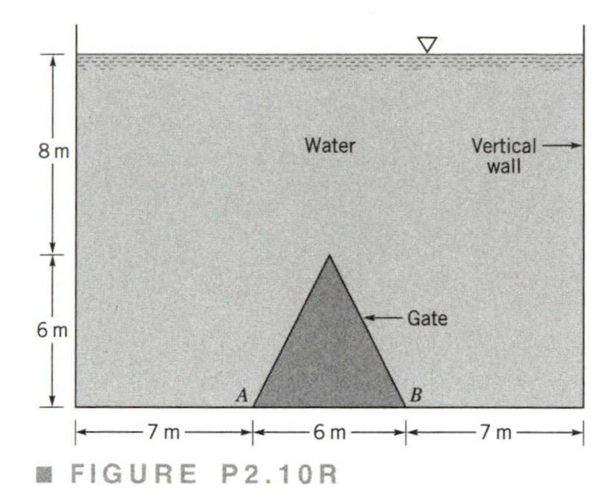

■ FIGURE P2.10R

$F_R = \gamma h_c A$ where $h_c = 8m + \frac{2}{3}(6m) = 12m$

Thus,

$$F_R = \left(9800 \frac{N}{m^3}\right)(12m)\left(\tfrac{1}{2}\right)(6m \times 6m) = 2120 \text{ kN}$$

To locate F_R,

$$y_R = \frac{I_{xc}}{y_c A} + y_c \quad \text{with } y_c = h_c \text{ so that}$$

$$y_R = \frac{\frac{1}{36}(6m)(6m)^3}{(12m)(\frac{1}{2})(6m)^2} + 12m = 12.167m$$

Thus, to determine the moment about AB

$$M_{AB} = (2120 \times 10^3 N)(14m - 12.167m)$$

$$= \underline{\underline{3890 \text{ kN}\cdot\text{m}}}$$

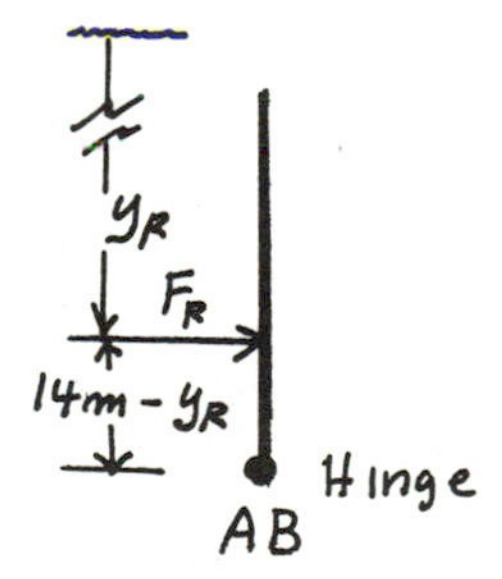

2.11R

2.11R (Force on plane surface) The rectangular gate CD of Fig P2.11R is 1.8 m wide and 2.0 m long. Assuming the material of the gate to be homogeneous and neglecting friction at the hinge C, determine the weight of the gate necessary to keep it shut until the water level rises to 2.0 m above the hinge.
(ANS: 180 kN)

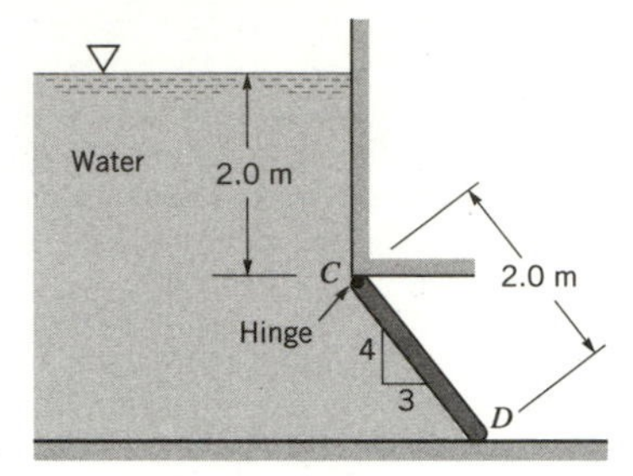

■ FIGURE P2.11R

$$F_R = \gamma h_c A$$

where $h_c = 2m + \frac{1}{2}\left[\left(\frac{4}{5}\right)(2m)\right] = 2.8m$

Thus,

$$F_R = \left(9.80 \frac{kN}{m^3}\right)(2.8\,m)(1.8\,m \times 2\,m)$$
$$= 98.8\ kN$$

Also,

$$y_R = \frac{I_{xc}}{y_c A} + y_c \qquad \text{where } y_c = \frac{2\,m}{\left(\frac{4}{5}\right)} + 1m = 3.5m$$

so that

$$y_R = \frac{\left(\frac{1}{12}\right)(1.8m)(2m)^3}{(3.5m)(1.8m \times 2m)} + 3.5m = 3.595m$$

For equilibrium,

$\sum M_o = 0$ (Note: Set $F_D = 0$ to obtain minimum weight)

and

$$W\left(\frac{1}{2}\right)\left[\left(\frac{3}{5}\right)(2m)\right] - F_R\left(y_R - \frac{2m}{\left(\frac{4}{5}\right)}\right) = 0$$

or

$$W = \frac{(98.8\,kN)(3.595m - 2.5m)}{\left(\frac{1}{2}\right)\left[\left(\frac{3}{5}\right)(2m)\right]} = \underline{\underline{180\,kN}}$$

2.12R

2.12R (Force on curved surface) A gate in the form of a partial cylindrical surface (called a *Tainter gate*) holds back water on top of a dam as shown in Fig. P2.12R. The radius of the surface is 22 ft, and its length is 36 ft. The gate can pivot about point *A*, and the pivot point is 10 ft above the seat, *C*. Determine the magnitude of the resultant water force on the gate. Will the resultant pass through the pivot? Explain.

(ANS: 118,000 lb)

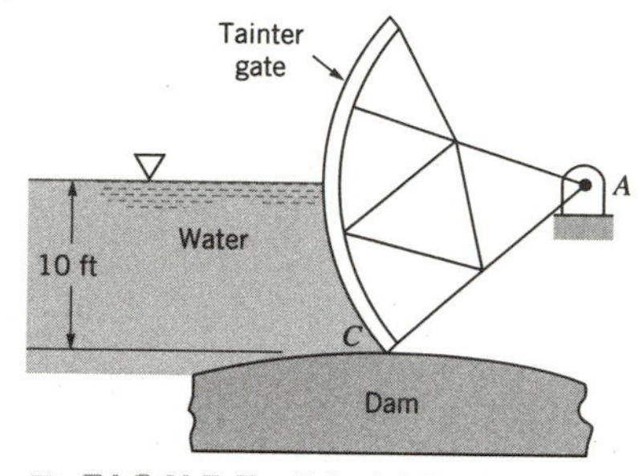

■ FIGURE P2.12R

Let F_G be force of gate on fluid

and

$$F_1 = \gamma h_c A$$

$$= \left(62.4\ \frac{\text{lb}}{\text{ft}^3}\right)\left(\frac{10\ \text{ft}}{2}\right)(10\ \text{ft} \times 36\ \text{ft})$$

$$= 112{,}000\ \text{lb}$$

(Note: All lengths in ft)

$$\sin\theta = \frac{10}{22}$$

$$\therefore\ \theta = 27.0°$$

$$\ell_{DE} = \ell_{BC} = 22 - \ell_{AB}$$

$$= 22 - 22\cos\theta$$

$$= 2.40\ \text{ft}$$

Also,

$$F_2 = \left(62.4\ \frac{\text{lb}}{\text{ft}^3}\right)(10\ \text{ft})(\ell_{DE} \times 36\ \text{ft})$$

$$= \left(62.4\ \frac{\text{lb}}{\text{ft}^3}\right)(10\ \text{ft})(2.40\ \text{ft} \times 36\ \text{ft})$$

$$= 53{,}900\ \text{lb}$$

and

$$\mathcal{W} = \gamma \forall_{CDE} = \gamma(A_{CDE} \times 36\ \text{ft})$$

where:

$$A_{CDE} = A_{BCDE} - A_{BCE} = (10 \times \ell_{DE}) - A_{BCE}$$

$$A_{BCE} = A_{ACE} - A_{ABE} = \pi\,(22\ \text{ft})^2\left(\frac{27.0°}{360°}\right) - \frac{1}{2}\left(10\ \text{ft} \times [22\cos 27.0°]\ \text{ft}\right)$$

$$= 16.0\ \text{ft}^2$$

Thus,

$$A_{CDE} = (10\ \text{ft} \times 2.40\ \text{ft}) - 16.0\ \text{ft}^2 = 8.00\ \text{ft}^2$$

and

$$\mathcal{W} = \left(62.4\ \frac{\text{lb}}{\text{ft}^3}\right)(8.00\ \text{ft}^2)(36\ \text{ft}) = 18{,}000\ \text{lb}$$

continued

For equilibrium,

$$\Sigma F_x = 0$$

or

$$F_{Gx} = F_1 = 112{,}000 \text{ lb}$$

Also,

$$\Sigma F_y = 0$$

or

$$F_{Gy} = F_2 - W = 53{,}900 \text{ lb} - 18{,}000 \text{ lb} = 35{,}900 \text{ lb}$$

Thus,

$$F_G = \sqrt{(F_{Gx})^2 + (F_{Gy})^2} = \sqrt{(112{,}000 \text{ lb})^2 + (35{,}900 \text{ lb})^2}$$

$$= \underline{\underline{118{,}000 \text{ lb}}}$$

The direction of all differential forces acting on the gate is perpendicular to the gate surface, and therefore, the resultant must pass through the intersection of all these forces which is at point A. <u>Yes</u>.

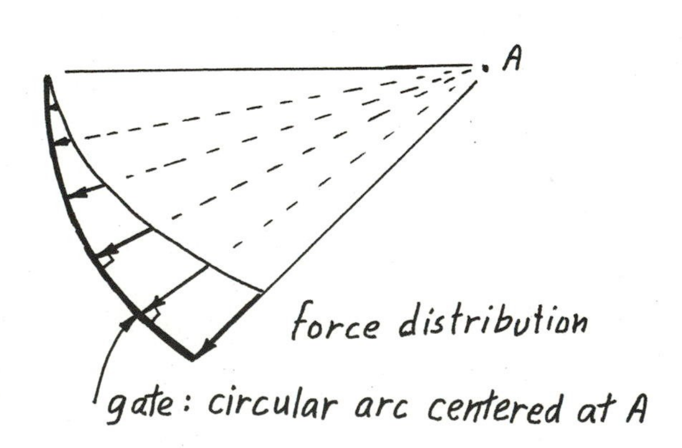

2.13 R

2.13R (Force on curved surface) A conical plug is located in the side of a tank as shown in Fig. 2.13R. **(a)** Show that the horizontal component of the force of the water on the plug does not depend on h. **(b)** For the depth indicated, what is the magnitude of this component?

(ANS: 735 lb)

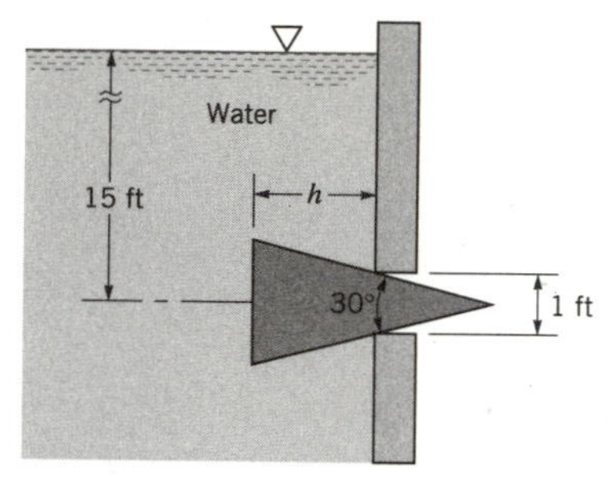

■ FIGURE P2.13R

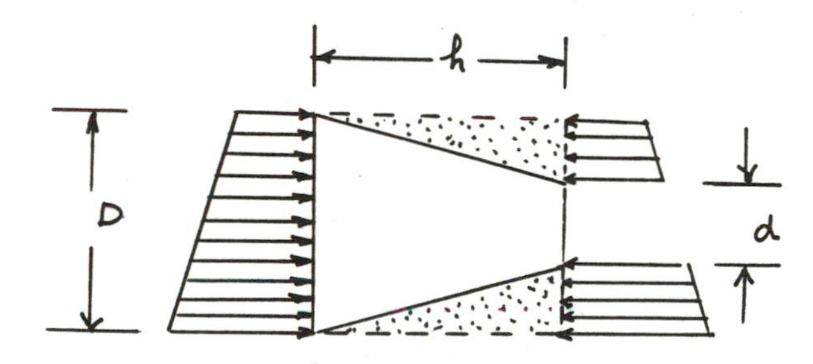

(a) Consider a cylinder of fluid of diameter, D, and length, h, with the plug removed (see figure). The pressure distributions over the right and left surfaces are shown. We note that the pressures cancel except for the center area of diameter, d. The pressure distribution over this center area will yield a resultant which is independent of h and depends only on the fluid specific weight, the fluid depth, and the hole diameter, d.

(b) For a circular area of diameter, $d = 1 ft$,

$$F_R = \gamma h_c A$$

$$= \left(62.4 \frac{lb}{ft^3}\right)(15\ ft)\left(\frac{\pi}{4}\right)(1\ ft)^2 = \underline{\underline{735\ lb}}$$

2.14R

2.14R (Force on curved surface) The 9-ft-long cylinder of Fig. P2.14R floats in oil and rests against a wall. Determine the horizontal force the cylinder exerts on the wall at the point of contact, *A*.

(ANS: 2300 lb)

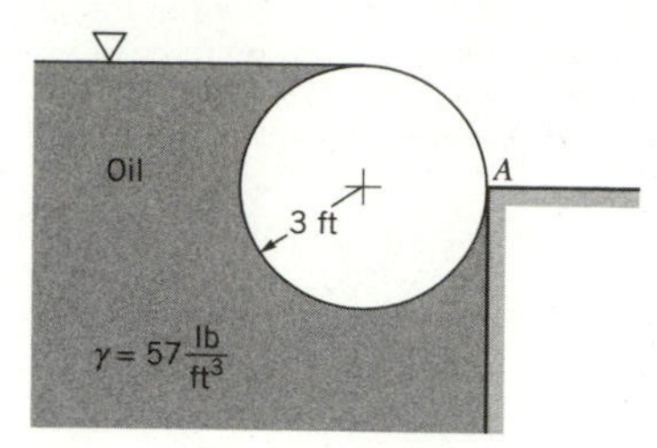

■ FIGURE P2.14R

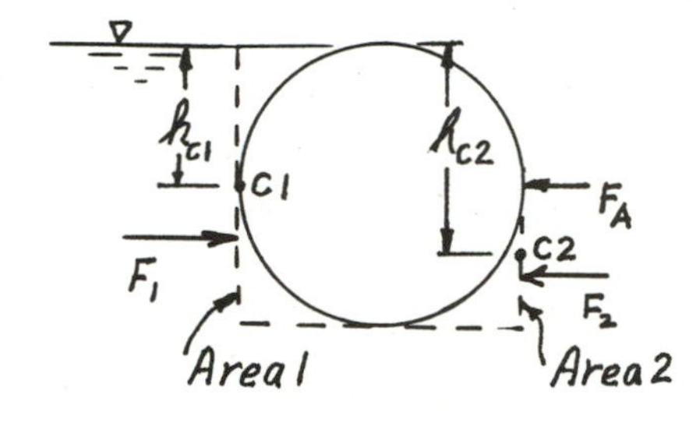

The horizontal forces acting on the free-body-diagram are shown on the figure. For equilibrium,

$$F_A = F_1 - F_2$$

where F_A is the horizontal force the wall exerts on the cylinder.

Since,

$$F_1 = \gamma h_{c1} A_1$$

$$= \left(57.0 \frac{lb}{ft^3}\right)\left(\frac{6 ft}{2}\right)(6 ft \times 9 ft)$$

$$= 9230 \ lb$$

and

$$F_2 = \gamma h_{c2} A_2$$

$$= \left(57.0 \frac{lb}{ft^3}\right)\left(3 ft + \frac{3}{2} ft\right)(3 ft \times 9 ft)$$

$$= 6930 \ lb$$

then

$$F_A = 9230 \ lb - 6930 \ lb = \underline{\underline{2300 \ lb \longrightarrow \text{on the wall}}}$$

2.15R

2.15R (Buoyancy) A hot-air balloon weighs 500 lb, including the weight of the balloon, the basket, and one person. The air outside the balloon has a temperature of 80 °F, and the heated air inside the balloon has a temperature of 150 °F. Assume the inside and outside air to be at standard atmospheric pressure of 14.7 psia. Determine the required volume of the balloon to support the weight. If the balloon had a spherical shape, what would be the required diameter?

(ANS: 59,200 ft^3; 48.3 ft)

Volume = $\forall$

$\gamma_{\substack{inside\\air}}$ F_B $\mathcal{W}_a$ $\gamma_{\substack{outside\\air}}$ $\mathcal{W}_b$

For equilibrium,

$$\Sigma F_{vertical} = 0$$

so that

$$F_B = \mathcal{W}_a + \mathcal{W}_b$$

where:

F_B = buoyant force

$\mathcal{W}_a$ = weight of air inside balloon

$\mathcal{W}_b$ = weight of basket and load

Thus,

$$(\gamma_{\substack{outside\\air}})\forall = (\gamma_{\substack{inside\\air}})\forall + \mathcal{W}_b \qquad (1)$$

From the ideal gas law $p = \rho RT = \frac{\gamma}{g} RT$ or

$$\gamma = \frac{g\,p}{RT}$$

For outside air with $T = 80°F + 460 = 540°R$,

$$\gamma_{\substack{outside\\air}} = \frac{\left(32.2\,\frac{ft}{s^2}\right)\left(14.7\,\frac{lb}{in.^2}\right)\left(144\,\frac{in.^2}{ft^2}\right)}{\left(1716\,\frac{ft\cdot lb}{slug\cdot °R}\right)(540\,°R)} = 0.07356\,\frac{lb}{ft^3}$$

Similarly for inside air with $T = 150°F + 460 = 610°R$,

$$\gamma_{\substack{inside\\air}} = \left(\frac{540\,°R}{610\,°R}\right)\left(0.07356\,\frac{lb}{ft^3}\right) = 0.06512\,\frac{lb}{ft^3}$$

Thus, from Eq. (1)

$$\forall = \frac{\mathcal{W}_b}{\gamma_{\substack{outside\\air}} - \gamma_{\substack{inside\\air}}} = \frac{500\,lb}{0.07356\,\frac{lb}{ft^3} - 0.06512\,\frac{lb}{ft^3}} = \underline{\underline{59{,}200\,ft^3}}$$

For spherical shape (with d = diameter),

$$\frac{\pi}{6} d^3 = 59{,}200\,ft^3 \quad \text{so that} \quad \underline{\underline{d = 48.3\,ft}}$$

2.16R

2.16R (Buoyancy) An irregularly shaped piece of a solid material weighs 8.05 lb in air and 5.26 lb when completely submerged in water. Determine the density of the material.
(ANS: 5.60 $slugs/ft^3$)

W (in air) $= \rho g \times$ (volume) where $\rho \sim$ density of material

W (in water) $= \rho g \times$ (volume) $-$ buoyant force

Thus, $= \rho g \times$ (volume) $- \rho_{H_2O}\, g \times$ (volume)

$$\frac{W\ (\text{in air})}{W\ (\text{in water})} = \frac{\rho}{\rho - \rho_{H_2O}} = \frac{1}{1 - \frac{\rho_{H_2O}}{\rho}}$$

or

$$\rho = \frac{\rho_{H_2O}}{1 - \frac{W\ (\text{in water})}{W\ (\text{in air})}} = \frac{1.94\ \frac{slugs}{ft^3}}{1 - \frac{5.26\ lb}{8.05\ lb}} = \underline{\underline{5.60\ \frac{slugs}{ft^3}}}$$

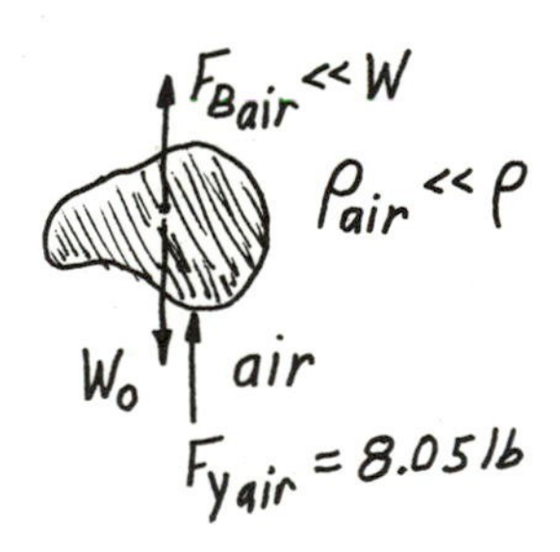

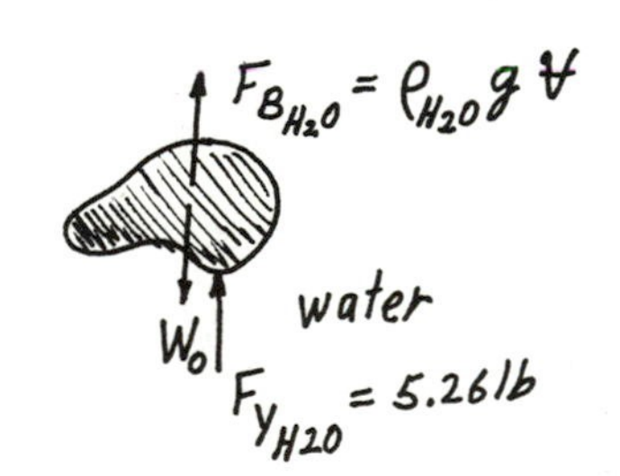

2.17R

2.17R (Buoyancy, force on plane surface) A cube, 4 ft on a side, weighs 3000 lb and floats half-submerged in an open tank as shown in Fig. P2.17R. For a liquid depth of 10 ft, determine the force of the liquid on the inclined section *AB* of the tank wall. The width of the wall is 8 ft. Show the magnitude, direction, and location of the force on a sketch.

(ANS: 75,000 lb on centerline, 13.33 ft along wall from free surface)

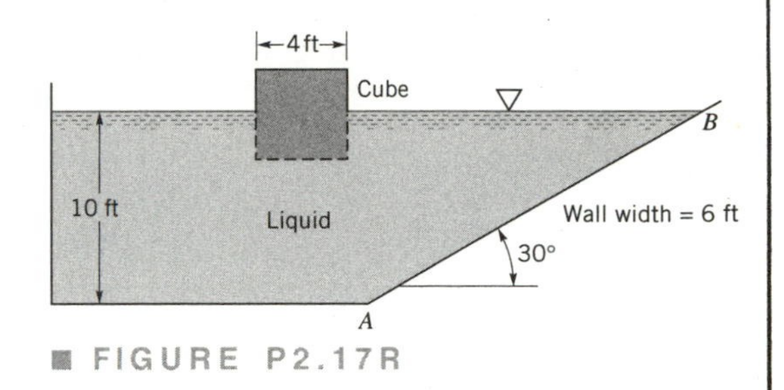

■ FIGURE P2.17R

Since the cube is floating,

$$\Sigma F_{vertical} = 0$$

or

$$W = F_{buoyant} = \gamma \times \text{Vol}$$

so that

$$\gamma = \frac{3000 \text{ lb}}{\frac{1}{2}(4\text{ft} \times 4\text{ft} \times 4\text{ft})} = 93.8 \frac{\text{lb}}{\text{ft}^3}$$

$F_{buoyant}$ / Cube / W

For the tank wall AB,

$$F_R = \gamma h_c A$$

Where $h_c = 5$ ft

so that

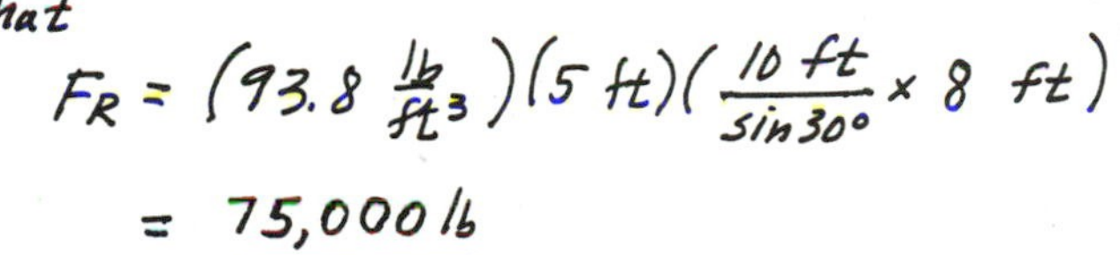

$$F_R = \left(93.8 \frac{\text{lb}}{\text{ft}^3}\right)(5 \text{ ft})\left(\frac{10 \text{ ft}}{\sin 30°} \times 8 \text{ ft}\right)$$

$$= \underline{\underline{75{,}000 \text{ lb}}}$$

F_R, h_c, B, c, y_R, CP, A

Also,

$$y_R = \frac{I_{xc}}{y_c A} + y_c \quad \text{where} \quad y_c = \frac{h_c}{\sin 30°} = \frac{5 \text{ ft}}{\sin 30°} = 10 \text{ ft}$$

and

$$y_R = \frac{\frac{1}{12}(8\text{ ft})(20\text{ ft})^3}{(10\text{ ft})(20\text{ft} \times 8\text{ ft})} + 10\text{ ft} = \underline{\underline{13.33 \text{ ft}}}$$

Thus,

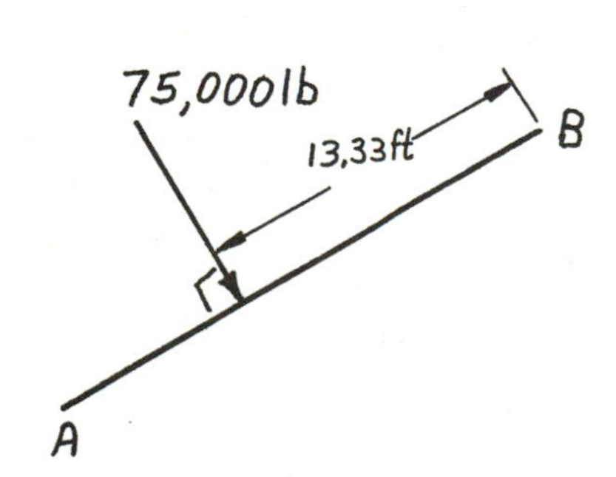

2.18R

2.18R (Rigid body motion) A container that is partially filled with water is pulled with a constant acceleration along a plane horizontal surface. With this acceleration the water surface slopes downward at an angle of 40° with respect to the horizontal. Determine the acceleration. Express your answer in m/s^2.

(ANS: $8.23\ m/s^2$)

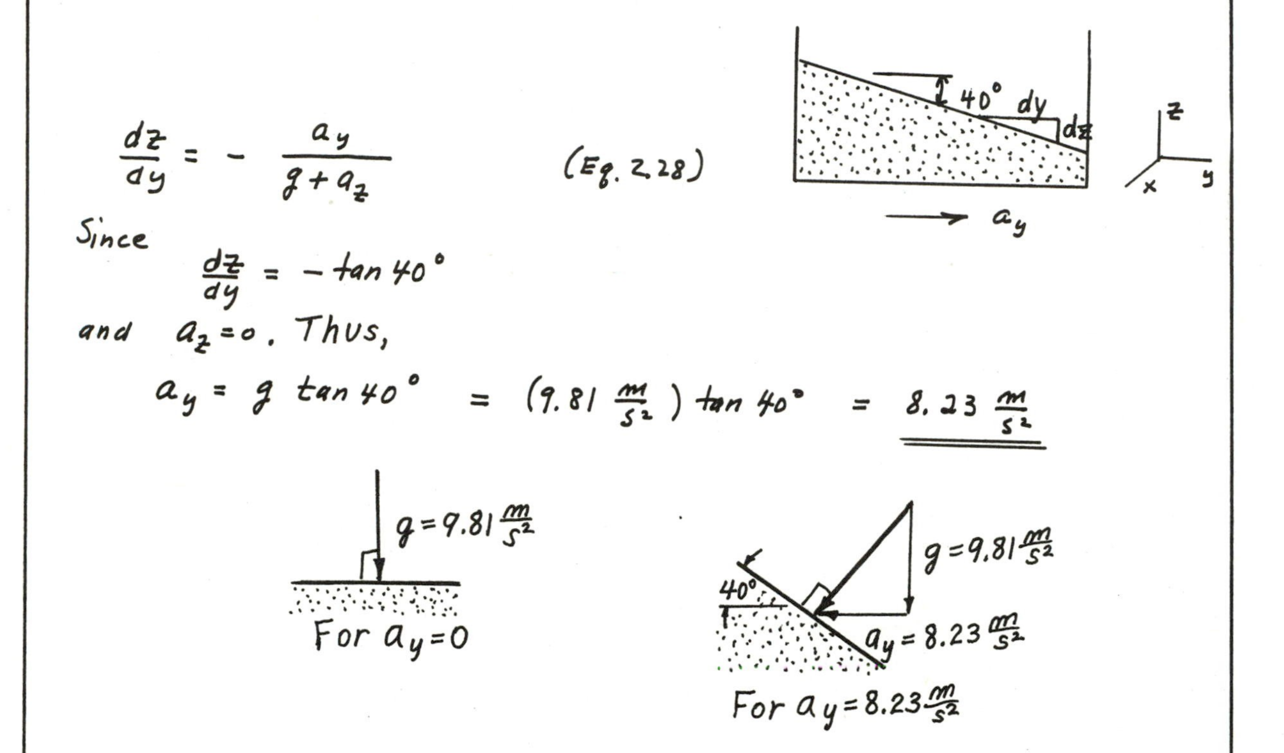

$$\frac{dz}{dy} = -\frac{a_y}{g + a_z} \qquad \text{(Eq. 2.28)}$$

Since

$$\frac{dz}{dy} = -\tan 40^\circ$$

and $a_z = 0$. Thus,

$$a_y = g \tan 40^\circ = \left(9.81 \frac{m}{s^2}\right) \tan 40^\circ = \underline{\underline{8.23 \frac{m}{s^2}}}$$

2.19R

2.19R (Rigid-body motion) An open, 2-ft-diameter tank contains water to a depth of 3 ft when at rest. If the tank is rotated about its vertical axis with an angular velocity of 160 rev/min, what is the minimum height of the tank walls to prevent water from spilling over the sides?
(ANS: 5.18 ft)

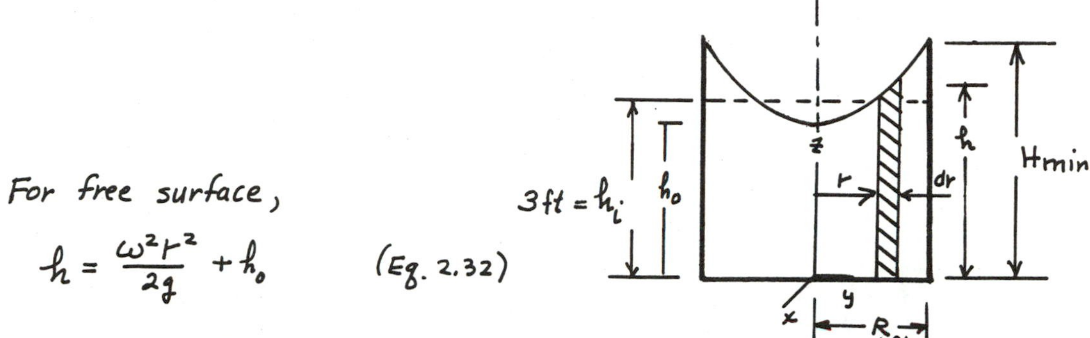

For free surface,

$$h = \frac{\omega^2 r^2}{2g} + h_o \qquad \text{(Eq. 2.32)}$$

The volume of fluid in the rotating tank is given by

$$\forall_f = \int_0^R 2\pi r h \, dr = 2\pi \int_0^R \left(\frac{\omega^2 r^3}{2g} + h_o r \right) dr$$

$$= \frac{\pi \omega^2 R^4}{4g} + \pi h_o R^2$$

$$= \frac{\pi \left(160 \frac{\text{rev}}{\text{min}} \times 2\pi \frac{\text{rad}}{\text{rev}} \times \frac{1 \text{ min}}{60 \text{ s}}\right)^2 (1 \text{ ft})^4}{4 \left(32.2 \frac{\text{ft}}{\text{s}^2}\right)} + \pi h_o (1 \text{ ft})^2$$

$$= \pi (2.18 + h_o) \text{ ft}^3 \qquad (\text{with } h_o \text{ in ft})$$

Since the initial volume,

$$\forall_i = \pi R^2 h_i = \pi (1 \text{ ft})^2 (3 \text{ ft}) = 3\pi \text{ ft}^3$$

and the final volume must be equal,

$$\forall_f = \forall_i$$

or

$$\pi (2.18 + h_o) \text{ ft}^3 = 3\pi \text{ ft}^3$$

and

$$h_o = 0.820 \text{ ft}$$

Thus, from the first equation (Eq. 2.32)

$$h = \frac{\omega^2 r^2}{2g} + 0.820 \text{ ft}$$

and

$$H_{min} = \frac{\left(160 \frac{\text{rev}}{\text{min}} \times 2\pi \frac{\text{rad}}{\text{rev}} \times \frac{1 \text{ min}}{60 \text{ s}}\right)^2 (1 \text{ ft})^2}{2 \left(32.2 \frac{\text{ft}}{\text{s}^2}\right)} + 0.820 \text{ ft} = \underline{\underline{5.18 \text{ ft}}}$$

3

Elementary Fluid Dynamics— The Bernoulli Equation

Flow past a blunt body: On any object placed in a moving fluid there is a stagnation point on the front of the object where the velocity is zero. This location has a relatively large pressure and divides the flow field into two portions—one flowing over the body, and one flowing under the body. (Dye in water.) (Photograph by B. R. Munson.)

3.1R

3.1R (F = ma along streamline) What pressure gradient along the streamline, dp/ds, is required to accelerate air at standard temperature and pressure in a horizontal pipe at a rate of 300 ft/s^2?

(ANS: $-0.714\ lb/ft^3$)

$$\frac{\partial p}{\partial s} = -\gamma \sin\theta - \rho V \frac{\partial V}{\partial s}$$ where $\theta = 0$ and $V\frac{\partial V}{\partial s} = a_s = 300\ \frac{ft}{s^2}$

Thus,

$$\frac{\partial p}{\partial s} = -\rho a_s = -2.38\times10^{-3}\ \frac{slug}{ft^3}\left(300\ \frac{ft}{s^2}\right) = \underline{\underline{-0.714\ \left(\frac{lb}{ft^2}\right)/ft}}$$

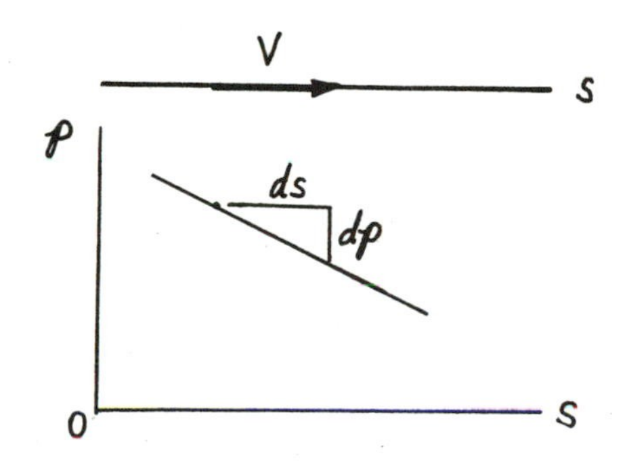

3.2R

3.2R (F = ma normal to streamline) An incompressible, inviscid fluid flows steadily with circular streamlines around a horizontal bend as shown in Fig. P3.2R. The radial variation of the velocity profile is given by $rV = r_0V_0$, where V_0 is the velocity at the inside of the bend which has radius $r = r_0$. Determine the pressure variation across the bend in terms of V_0, r_0, ρ, r, and p_0, where p_0 is the pressure at $r = r_0$. Plot the pressure distribution, $p = p(r)$, if $r_0 = 1.2$ m, $r_1 = 1.8$ m, $V_0 = 12$ m/s, $p_0 = 20$ kN/m^2, and the fluid is water. Neglect gravity.

(ANS: $p_0 + 0.5\rho V_0^2[1 - (r_0/r)^2]$)

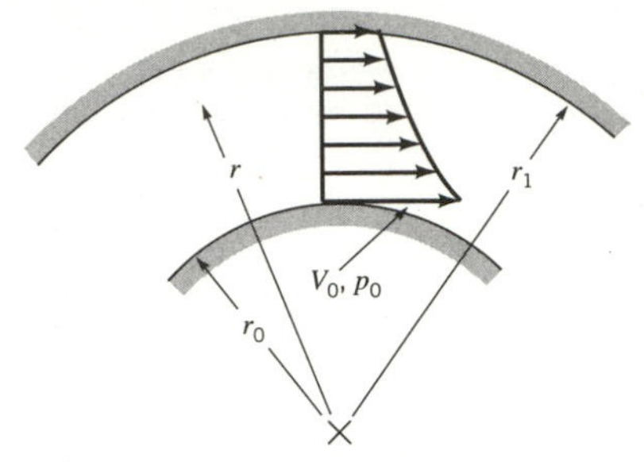

■ FIGURE P3.2R

$$-\gamma\frac{dz}{dn} - \frac{\partial p}{\partial n} = \frac{\rho V^2}{\mathcal{R}} \quad \text{with } \frac{dz}{dn} = 0 \text{ and } r = \mathcal{R} = r_1 - n, \text{ or } \frac{\partial}{\partial n} = -\frac{\partial}{\partial r}$$

Thus,

$$\frac{dp}{dr} = +\frac{\rho V^2}{r} \quad \text{or} \quad \int_{p_0}^{p} dp = +\rho\int_{r_0}^{r} \frac{V^2}{r}\,dr$$

But

$$\int_{r_0}^{r} \frac{V^2}{r}\,dr = \int_{r_0}^{r} \frac{(r_0V_0)^2}{r^3}\,dr = -\frac{r_0^2V_0^2}{2}\left[\frac{1}{r^2} - \frac{1}{r_0^2}\right]$$

Hence,

$$p = p_0 - \frac{1}{2}\rho r_0^2V_0^2\left[\frac{1}{r^2} - \frac{1}{r_0^2}\right] = \underline{\underline{p_0 + \frac{1}{2}\rho V_0^2\left[1 - \left(\frac{r_0}{r}\right)^2\right]}}$$

For the given data:

$$p = 20\frac{kN}{m^2} + \frac{1}{2}\left(999\frac{kg}{m^3}\right)\left(12\frac{m}{s}\right)^2\left[1 - \left(\frac{1.2}{r}\right)^2\right]$$

$$= 20 + 71.9\left[1 - \left(\frac{1.2}{r}\right)^2\right] \frac{kN}{m^2}, \text{ where } r \sim m$$

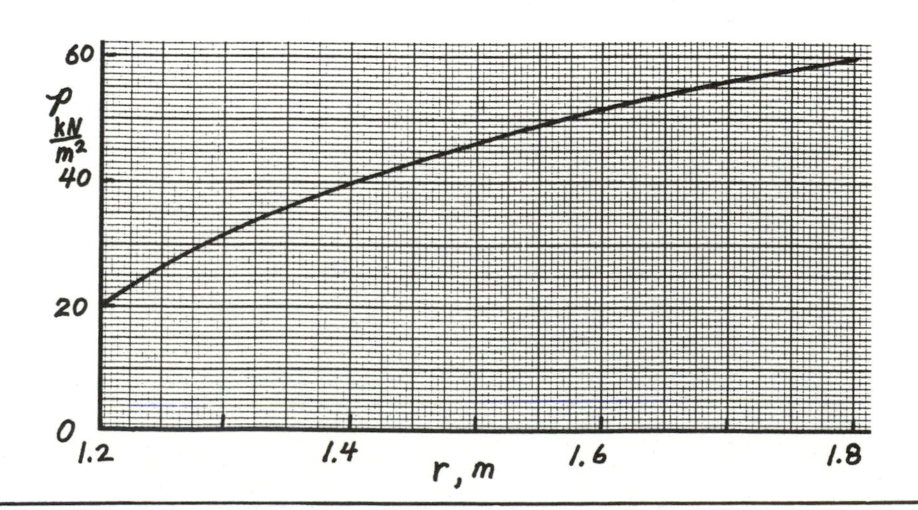

3.3R

3.3R (Stagnation pressure) A hang glider soars through standard sea level air with an airspeed of 10 m/s. What is the gage pressure at a stagnation point on the structure?

(ANS: 61.5 Pa)

$V_1 = 10\,\frac{m}{s}$

(1) $p_1 = 0$ (2) $V_2 = 0$

$$\frac{p_1}{\gamma} + \frac{V_1^2}{2g} + z_1 = \frac{p_2}{\gamma} + \frac{V_2^2}{2g} + z_2 \quad \text{with } z_1 = z_2,\ p_1 = 0,\ V_1 = 10\,\frac{m}{s} \text{ and } V_2 = 0$$

Thus, $p_2 = \frac{1}{2}\rho V_1^2 = \frac{1}{2}\left(1.23\,\frac{kg}{m^3}\right)\left(10\,\frac{m}{s}\right)^2 = 61.5\,\frac{kg\,m}{s^2}/m^2$

$= \underline{\underline{61.5\ Pa}}$

3.4R

3.4R (Bernoulli equation) The pressure in domestic water pipes is typically 60 psi above atmospheric. If viscous effects are neglected, determine the height reached by a jet of water through a small hole in the top of the pipe.

(ANS: 138 ft)

(2) $z_2 = h$

h

(1) $z_1 = 0$

$$\frac{p_1}{\gamma} + \frac{V_1^2}{2g} + z_1 = \frac{p_2}{\gamma} + \frac{V_2^2}{2g} + z_2$$

but $p_1 = 60\ psi$

$p_2 = 0$

$V_2 = 0$

$V_1 \approx 0$ if the hole diameter is much smaller than the pipe diameter

Thus,

$$\frac{p_1}{\gamma} = h$$

or

$$h = \frac{60\,\frac{lb}{in.^2}\left(144\,\frac{in.^2}{ft^2}\right)}{62.4\,\frac{lb}{ft^3}}$$

$= \underline{\underline{138\ ft}}$

Note: Because of viscous effects between the water and the pipe and the water and the air, the actual value of would be less than 138 ft.

3.5R

3.5R (Heads) A 4-in.-diameter pipe carries 300 gal/min of water at a pressure of 30 psi. Determine **(a)** the pressure head in feet of water, **(b)** the velocity head, and **(c)** the total head with reference to a datum plane 20 ft below the pipe.

(ANS: 69.2 ft; 0.909 ft; 90.1 ft)

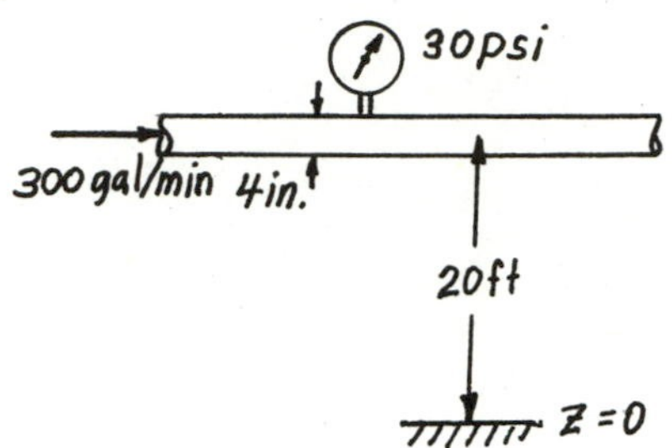

(a) $\frac{p}{\gamma} = \frac{30\,\frac{lb}{in.^2}\left(144\,\frac{in^2}{ft^2}\right)}{62.4\,\frac{lb}{ft^3}} = \underline{\underline{69.2 ft}}$

(b) $Q = 300\,\frac{gal}{min}\left(231\,\frac{in^3}{gal}\right)\left(\frac{1\,min}{60\,s}\right)\left(\frac{1\,ft^3}{1728\,in.^3}\right) = 0.668\,\frac{ft^3}{s}$

so that

$V = \frac{Q}{A} = \frac{0.668\,\frac{ft^3}{s}}{\frac{\pi}{4}\left(\frac{4}{12}ft\right)^2} = 7.65\,\frac{ft}{s}$

or

$\frac{V^2}{2g} = \frac{\left(7.65\,\frac{ft}{s}\right)^2}{2\left(32.2\,\frac{ft}{s^2}\right)} = \underline{\underline{0.909 ft}}$

(c) $\frac{p}{\gamma} + \frac{V^2}{2g} + z = 69.2 + 0.909 + 20 = \underline{\underline{90.1 ft}}$

3.6R

3.6R (Free jet) Water flows from a nozzle of triangular cross section as shown in Fig. P3.6R. After it has fallen a distance of 2.7 ft, its cross section is circular (because of surface tension effects) with a diameter $D = 0.11$ ft. Determine the flowrate, Q.

(ANS: 0.158 ft^3/s)

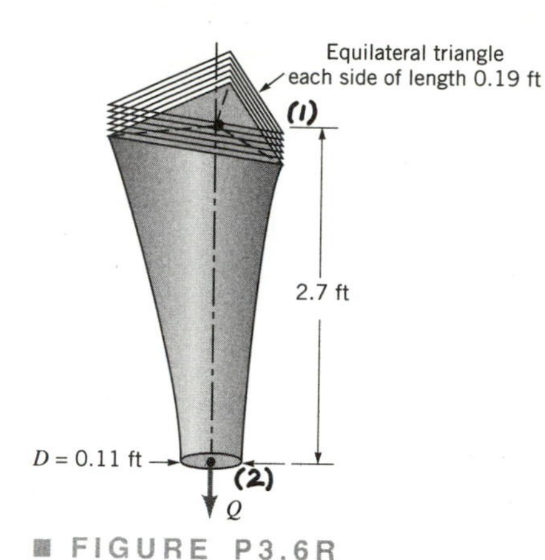

FIGURE P3.6R

$$\frac{p_1}{\gamma} + \frac{V_1^2}{2g} + z_1 = \frac{p_2}{\gamma} + \frac{V_2^2}{2g} + z_2$$

where $p_1 = p_2 = 0$, $z_2 = 0$, $z_1 = 2.7\,ft$

and

$$V_1 = \frac{Q}{A_1} \quad , \quad V_2 = \frac{Q}{A_2}$$

Thus,

$$\left(\frac{Q}{A_1}\right)^2 + 2gz_1 = \left(\frac{Q}{A_2}\right)^2 \quad \text{or} \quad Q = \left[\frac{2gz_1}{\left(\frac{1}{A_2^2} - \frac{1}{A_1^2}\right)}\right]^{1/2} = \frac{A_2\sqrt{2gz_1}}{\sqrt{1 - (A_2/A_1)^2}}$$

but

$$A_2 = \frac{\pi}{4}(0.11\,ft)^2 = 0.00950\,ft^2$$

and

$$A_1 = \frac{1}{2}(0.19\,ft)(0.1645\,ft) = 0.0156\,ft^2$$

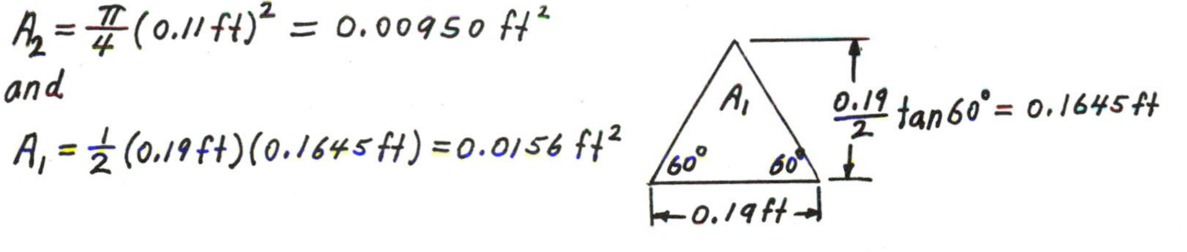

Thus,

$$Q = \frac{(0.00950\,ft^2)\sqrt{2(32.2\,\frac{ft}{s^2})(2.7\,ft)}}{\left[1 - \frac{(0.00950\,ft^2)^2}{(0.0156\,ft^2)^2}\right]^{1/2}} = \underline{\underline{0.158\,\frac{ft^3}{s}}}$$

3.7R

3.7R (Bernoulli/continuity) Water flows into a large tank at a rate of $0.011\ m^3/s$ as shown in Fig. P3.7R. The water leaves the tank through 20 holes in the bottom of the tank, each of which produces a stream of 10-mm diameter. Determine the equilibrium height, h, for steady state operation.

(ANS: 2.50 m)

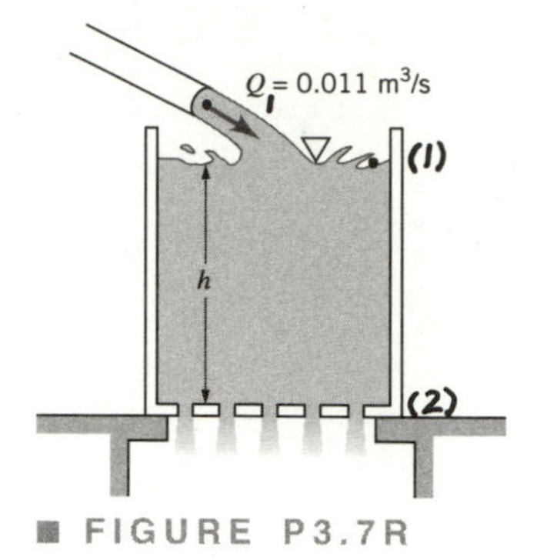

■ FIGURE P3.7R

$Q_1 = Q_2$ where $Q_1 = 0.011\ \frac{m^3}{s}$

and

$Q_2 = 20\, A_2 V_2 = 20 \frac{\pi}{4} D_2^2 V_2$

but

$\frac{p_1}{\gamma} + \frac{V_1^2}{2g} + z_1 = \frac{p_2}{\gamma} + \frac{V_2^2}{2g} + z_2$ where $p_1 = p_2 = 0$, $V_1 = 0$, and $z_1 - z_2 = h$

Thus,

$V_2 = \sqrt{2gh}$

so that

$0.011\ \frac{m^3}{s} = 20 \frac{\pi}{4} (0.01\,m)^2 \sqrt{2 (9.81\ \frac{m}{s^2}) h}$

or

$h = \underline{\underline{2.50\ m}}$

3.8R

3.8R (Bernoulli/continuity) Gasoline flows from a 0.3-m-diameter pipe in which the pressure is 300 kPa into a 0.15-m-diameter pipe in which the pressure is 120 kPa. If the pipes are horizontal and viscous effects are negligible, determine the flowrate.

(ANS: 0.420 m^3/s)

0.3 m; 0.15 m; (1); (2); $p_1 = 300\ kPa$; $p_2 = 120\ kPa$

$$\frac{p_1}{\gamma} + \frac{V_1^2}{2g} + z_1 = \frac{p_2}{\gamma} + \frac{V_2^2}{2g} + z_2 \quad \text{with } z_1 = z_2 \text{ and } \gamma = 6.67\ \frac{kN}{m^3}$$

Also, $A_1 V_1 = A_2 V_2$ or $V_2 = \left(\frac{D_1}{D_2}\right)^2 V_1 = \left(\frac{0.3\,m}{0.15\,m}\right)^2 V_1 = 4V_1$

Thus,

$$\frac{p_1}{\gamma} + \frac{V_1^2}{2g} = \frac{p_2}{\gamma} + \frac{16 V_1^2}{2g} \quad \text{or} \quad 15 V_1^2 = 2g\,\frac{p_1 - p_2}{\gamma}$$

so that

$$V_1 = \left[\frac{2\left(9.81\frac{m}{s^2}\right)(300\,kPa - 120\,kPa)}{15\left(6.67\frac{kN}{m^3}\right)}\right]^{1/2} = 5.94\ \frac{m}{s}$$

Thus,

$$Q = A_1 V_1 = \frac{\pi}{4}(0.3\,m)^2\left(5.94\ \frac{m}{s}\right) = \underline{\underline{0.420\ \frac{m^3}{s}}}$$

3.9R

3.9R (Bernoulli/continuity) Water flows steadily through the pipe shown in Fig. P3.9R such that the pressures at sections (1) and (2) are 300 kPa and 100 kPa, respectively. Determine the diameter of the pipe at section (2), D_2, if the velocity at section 1 is 20 m/s and viscous effects are negligible.

(ANS: 0.0688 m)

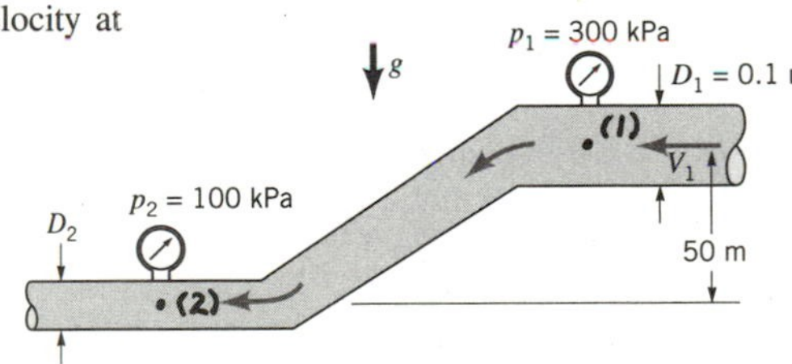

■ FIGURE P3.9R

$$\frac{p_1}{\gamma} + \frac{V_1^2}{2g} + z_1 = \frac{p_2}{\gamma} + \frac{V_2^2}{2g} + z_2$$

or

$$\frac{300 \times 10^3\,N/m^2}{9.80 \times 10^3\,N/m^3} + \frac{(20\,m/s)^2}{2(9.81\,m/s^2)} + 50\,m = \frac{100 \times 10^3\,N/m^2}{9.80 \times 10^3\,N/m^3} + \frac{V_2^2}{2(9.81\,m/s^2)}$$

Thus, $V_2 = 42.2\ m/s$

so that since $V_1 A_1 = V_2 A_2$ or $V_1 \frac{\pi}{4} D_1^2 = V_2 \frac{\pi}{4} D_2^2$, then

$$D_2 = \left[\frac{V_1}{V_2}\right]^{1/2} D_1 = \left[\frac{20\,m/s}{42.2\,m/s}\right]^{1/2}(0.1\,m) = \underline{\underline{0.0688\ m}}$$

3.10R

3.10R (Bernoulli/continuity) Water flows steadily through a diverging tube as shown in Fig. P3.10R. Determine the velocity, V, at the exit of the tube if frictional effects are negligible.

(ANS: 1.04 ft/s)

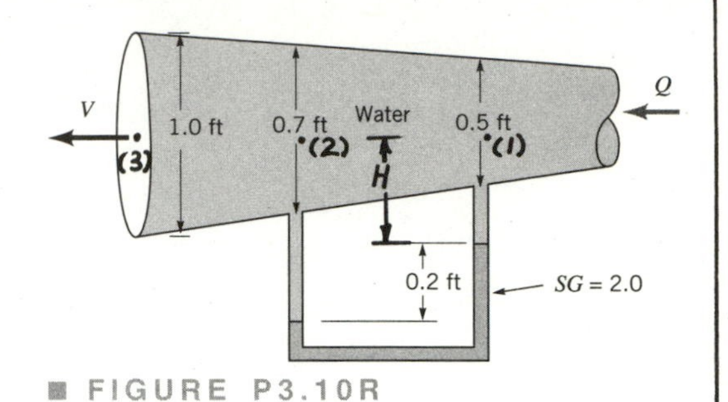

■ FIGURE P3.10R

$$\frac{p_1}{\gamma} + \frac{V_1^2}{2g} + z_1 = \frac{p_2}{\gamma} + \frac{V_2^2}{2g} + z_2 \qquad (z_1 = z_2) \qquad (1)$$

where $A_1 V_1 = A_2 V_2$, or

$$V_1 = \left(\frac{D_2}{D_1}\right)^2 V_2 = \left(\frac{0.7\,m}{0.5\,m}\right)^2 V_2 = 1.96\,V_2 \qquad (2)$$

Also, $p_2 + \gamma(H + 0.2\text{ft}) - SG\gamma(0.2\text{ft}) - \gamma H = p_1$

or

$$p_2 = p_1 + 2\gamma(0.2) - \gamma(0.2) = p_1 + 62.4\frac{lb}{ft^3}[2-1](0.2\,ft)$$

$$= p_1 + 12.48\frac{lb}{ft^2} \qquad (3)$$

By combining (1), (2), and (3):

$$\frac{p_1}{\gamma} + \frac{(1.96\,V_2)^2\,ft^2/s^2}{2(32.2\,ft/s^2)} = \frac{p_1}{\gamma} + \frac{12.48\,lb/ft^2}{62.4\,lb/ft^3} + \frac{V_2^2\,ft^2/s^2}{2(32.2\,ft/s^2)}$$

or $V_2 = 2.13\,ft/s$ and $V_1 = 1.96(2.13\,ft/s) = 4.17\,ft/s$

Thus, since $V_3 A_3 = V_2 A_2$, then

$$V_3 = \left(\frac{D_2}{D_3}\right)^2 V_2 = \left(\frac{0.7\,ft}{1\,ft}\right)^2 (2.13\,ft/s) = \underline{\underline{1.04\,ft/s}}$$

3.11R

3.11R (Bernoulli/continuity/Pitot tube) Two Pitot tubes and two static pressure taps are placed in the pipe contraction shown in Fig. P3.11R. The flowing fluid is water, and viscous effects are negligible. Determine the two manometer readings, h and H.

(ANS: 0; 0.252 ft)

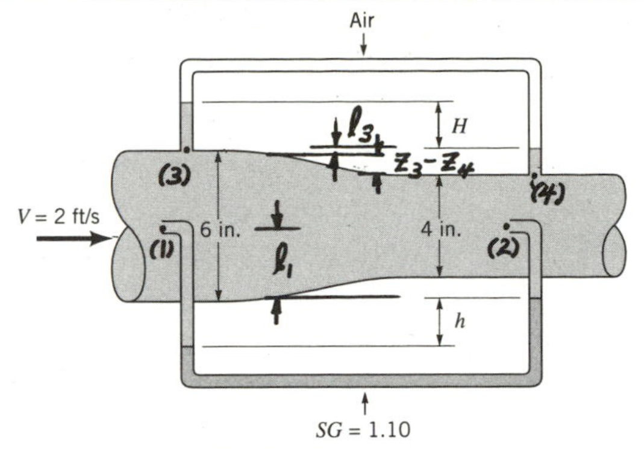

■ FIGURE P3.11R

$$\frac{p_1}{\gamma} + \frac{V_1^2}{2g} + z_1 = \frac{p_2}{\gamma} + \frac{V_2^2}{2g} + z_2 \quad \text{with } z_1 = z_2 \text{ and } V_1 = V_2 = 0$$

Thus, $p_1 = p_2$ so that manometer considerations give

$$p_1 + \gamma(\ell_1 + h) = p_2 + \gamma\ell_1 + 1.10\gamma h, \text{ or with } p_1 = p_2 \text{ this gives } \underline{\underline{h = 0}}$$

Also,

$$\frac{p_3}{\gamma} + \frac{V_3^2}{2g} + z_3 = \frac{p_4}{\gamma} + \frac{V_4^2}{2g} + z_4 \quad \text{where } z_3 = \frac{3}{12}\text{ ft},\ z_4 = \frac{2}{12}\text{ ft},\ V_3 = 2\frac{\text{ft}}{\text{s}},$$

and $A_3 V_3 = A_4 V_4$ or

$$V_4 = \frac{A_3}{A_4} V_3 = \frac{\frac{\pi}{4} D_3^2}{\frac{\pi}{4} D_4^2} V_3 = \left(\frac{6\text{ in.}}{4\text{ in.}}\right)^2 \left(2\frac{\text{ft}}{\text{s}}\right) = 4.50\frac{\text{ft}}{\text{s}}$$

Thus,
$$\frac{p_3 - p_4}{\gamma} = \left(\frac{2}{12} - \frac{3}{12}\right)\text{ft} + \frac{\left(4.50\frac{\text{ft}}{\text{s}}\right)^2 - \left(2\frac{\text{ft}}{\text{s}}\right)^2}{2\left(32.2\frac{\text{ft}}{\text{s}^2}\right)} = 0.169\text{ ft} \qquad (1)$$

But $p_3 - \gamma\ell_3 - \gamma H = p_4 - \gamma(z_3 - z_4 + \ell_3)$

or
$$\frac{p_3 - p_4}{\gamma} = H - (z_3 - z_4) \qquad (2)$$

From Eqs.(1) and (2) we obtain

$$0.169\text{ ft} = H - \left(\frac{3-2}{12}\text{ ft}\right) \quad \text{or} \quad \underline{\underline{H = 0.252\text{ ft}}}$$

3.12R

3.12R (Bernoulli/continuity) Water collects in the bottom of a rectangular oil tank as shown in Fig. P3.12R. How long will it take for the water to drain from the tank through a 0.02-m-diameter drain hole in the bottom of the tank? Assume quasi-steady flow.

(ANS: 2.45 hr)

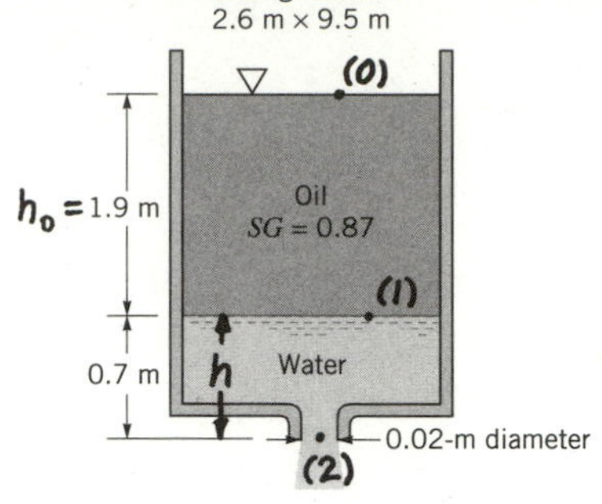

FIGURE P3.12R

$$\frac{p_1}{\gamma} + \frac{V_1^2}{2g} + z_1 = \frac{p_2}{\gamma} + \frac{V_2^2}{2g} + z_2$$ where $p_1 = p_0 + \gamma_0 h_0$, $p_0 = 0$, $p_2 = 0$, $z_1 = h$, $z_2 = 0$, and $V_1 = 0$

Thus, since $\frac{\gamma_0}{\gamma} = SG = 0.87$,

$$\frac{\gamma_0 h_0}{\gamma} + h = \frac{V_2^2}{2g} \quad \text{or} \quad V_2 = \sqrt{2g(h + SG h_0)}$$

or

$$V_2 = \sqrt{2\left(9.81\,\frac{m}{s^2}\right)(h + 0.87(1.9\,m))} = 4.43\sqrt{h + 1.653}\ \frac{m}{s},$$

where $h \sim m$

Also,

$$Q = A_2 V_2 = \frac{\pi}{4} D_2^2 V_2 = \frac{\pi}{4}(0.02\,m)^2\left[4.43\sqrt{h + 1.653}\ \frac{m}{s}\right] \tag{1}$$

$$= 1.39 \times 10^{-3}\sqrt{h + 1.653}\ \frac{m^3}{s}$$

and

$$Q = A_1\left(-\frac{dh}{dt}\right),$$

where

$$A_1 = 2.6\,m\,(9.5\,m) = 24.7\,m^2$$ Hence,

$$Q = -24.7\,\frac{dh}{dt} \tag{2}$$

Combine Eqs. (1) and (2) to give

$$\frac{dh}{dt} = -5.63 \times 10^{-5}\sqrt{h + 1.653}$$

or

$$\int_{h=0.7}^{h=0} \frac{dh}{\sqrt{h + 1.653}} = -5.63 \times 10^{-5}\int_{t=0}^{t=t_f} dt$$ where t_f = time to drain the water

or

$$2\sqrt{h + 1.653}\,\Big|_{0.7}^{0} = -5.63 \times 10^{-5}\, t_f$$

Thus,

$$t_f = 8.83 \times 10^3\,s = \underline{\underline{2.45\,hr}}$$

3.13R

3.13R (Cavitation) Water flows past the hydrofoil shown in Fig. P3.13R with an upstream velocity of V_0. A more advanced analysis indicates that the maximum velocity of the water in the entire flow field occurs at point B and is equal to $1.1V_0$. Calculate the velocity, V_0, at which cavitation will begin if the atmospheric pressure is 101 kPa (abs) and the vapor pressure of the water is 3.2 kPa (abs).

(ANS: 31.4 m/s)

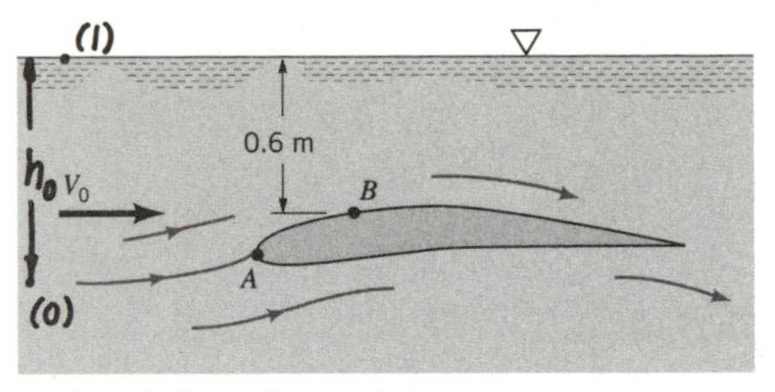

■ FIGURE P3.13R

$$\frac{p_0}{\gamma} + \frac{V_0^2}{2g} + z_0 = \frac{p_B}{\gamma} + \frac{V_B^2}{2g} + z_B \quad \text{where } p_B = p_v = 3.2 \text{ kPa (abs)}$$

$$p_0 = p_1 + \gamma h_0, \; z_0 = -h_0,$$

Thus, and $z_B = -0.6\text{m}$

$$\frac{p_1 + \gamma h_0}{\gamma} + \frac{V_0^2}{2g} - h_0 = \frac{p_v}{\gamma} + \frac{V_B^2}{2g} + z_B \qquad (1)$$

But $V_B = 1.1V_0$ and $p_1 = 101 \text{ kPa (abs)}$ so that Eq.(1) gives

$$\frac{101 \frac{kN}{m^2}}{9.80 \frac{kN}{m^3}} + \frac{V_0^2}{2(9.81\frac{m}{s^2})} = \frac{3.2 \frac{kN}{m^2}}{9.80\frac{kN}{m^3}} + \frac{(1.1V_0)^2}{2(9.81\frac{m}{s^2})} - 0.6\text{m}$$

or

$$V_0 = \underline{\underline{31.4 \frac{m}{s}}}$$

3.14R

3.14R (Flowrate) Water flows through the pipe contraction shown in Fig. P3.14R. For the given 0.2-m difference in manometer level, determine the flowrate as a function of the diameter of the small pipe, D.

(ANS: $0.0156\ m^3/s$)

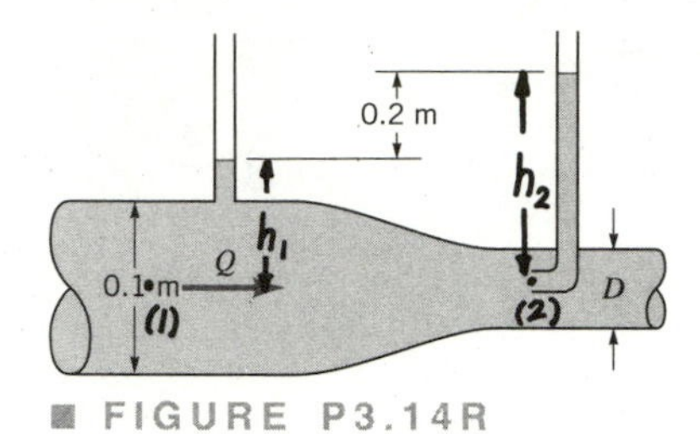

■ FIGURE P3.14R

$$\frac{p_1}{\gamma} + \frac{V_1^2}{2g} + z_1 = \frac{p_2}{\gamma} + \frac{V_2^2}{2g} + z_2, \quad \text{where } z_1 = z_2 \text{ and } V_2 = 0$$

Thus,

$$V_1 = \sqrt{2g\frac{(p_2 - p_1)}{\gamma}}$$

But

$$p_1 = \gamma h_1 \text{ and } p_2 = \gamma h_2, \text{ so that } p_2 - p_1 = \gamma(h_2 - h_1) = 0.2\gamma$$

Thus,

$$V_1 = \sqrt{2g\frac{0.2\gamma}{\gamma}} = \sqrt{2g(0.2)}$$

or

$$Q = A_1 V_1 = \frac{\pi}{4}(0.1m)^2\sqrt{2(9.81\tfrac{m}{s^2})(0.2m)} = \underline{\underline{0.0156\tfrac{m^3}{s}}}$$

3.15R

3.15R (Channel flow) Water flows down the ramp shown in the channel of Fig. P3.15R. The channel width decreases from 15 ft at section (1) to 9 ft at section (2). For the conditions shown, determine the flowrate.

(ANS: 509 ft^3/s)

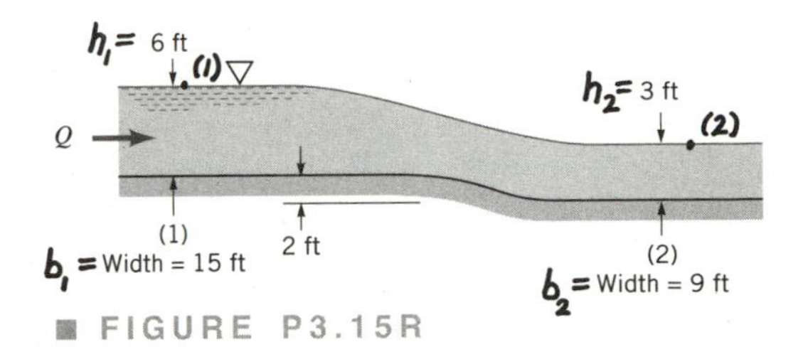

■ FIGURE P3.15R

$$\frac{p_1}{\gamma} + \frac{V_1^2}{2g} + z_1 = \frac{p_2}{\gamma} + \frac{V_2^2}{2g} + z_2 \qquad (1)$$

where $p_1 = 0$, $p_2 = 0$, $z_2 = 3\,ft$, and $z_1 = (6+2)\,ft = 8\,ft$

Also, $A_1 V_1 = A_2 V_2$

or

$$V_2 = \frac{h_1 b_1}{h_2 b_2} V_1 = \frac{(6\,ft)(15\,ft)}{(3\,ft)(9\,ft)} V_1 = 3.33\, V_1$$

Thus, Eq. (1) becomes

$$\left[3.33^2 - 1\right] V_1^2 = 2\left(32.2\,\frac{ft}{s^2}\right)(8-3)\,ft, \quad \text{or } V_1 = 5.65\,\frac{ft}{s}$$

Hence,

$$Q = A_1 V_1 = (6\,ft)(15\,ft)\left(5.65\,\frac{ft}{s}\right) = \underline{\underline{509\,\frac{ft^3}{s}}}$$

3.16R

3.16R (Channel flow) Water flows over the spillway shown in Fig. P3.16R. If the velocity is uniform at sections (1) and (2) and viscous effects are negligible, determine the flowrate per unit width of the spillway.

(ANS: 7.44 m^2/s)

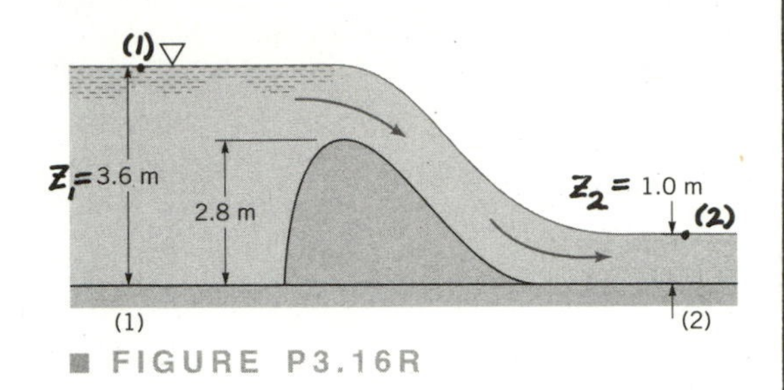

■ FIGURE P3.16R

$$\frac{p_1}{\gamma} + \frac{V_1^2}{2g} + z_1 = \frac{p_2}{\gamma} + \frac{V_2^2}{2g} + z_2$$

where, if points (1) and (2) are located on the free surface, $p_1 = 0$, $p_2 = 0$, $z_1 = 3.6\,m$, and $z_2 = 1.0\,m$.

Also, $A_1 V_1 = A_2 V_2$

or

$$V_1 = \frac{z_2}{z_1} V_2 = \frac{1.0\,m}{3.6\,m} V_2 = 0.278\, V_2$$

Thus, Eq. (1) becomes

$$\frac{V_2^2}{2(9.81\,\frac{m}{s^2})}\left[1 - (0.278)^2\right] = 3.6\,m - 1.0\,m, \quad \text{or} \quad V_2 = 7.44\,\frac{m}{s}$$

Hence,

$$q = V_2 z_2 = (7.44\,\tfrac{m}{s})(1.0\,m) = \underline{\underline{7.44\,\frac{m^2}{s}}}$$

3.17R

3.17R (Energy line/hydraulic grade line) Draw the energy line and hydraulic grade line for the flow shown in Problem 3.43.

3.43 A smooth plastic, 10-m-long garden hose with an inside diameter of 20 mm is used to drain a wading pool as is shown in Fig. P3.43. If viscous effects are neglected, what is the flowrate from the pool?

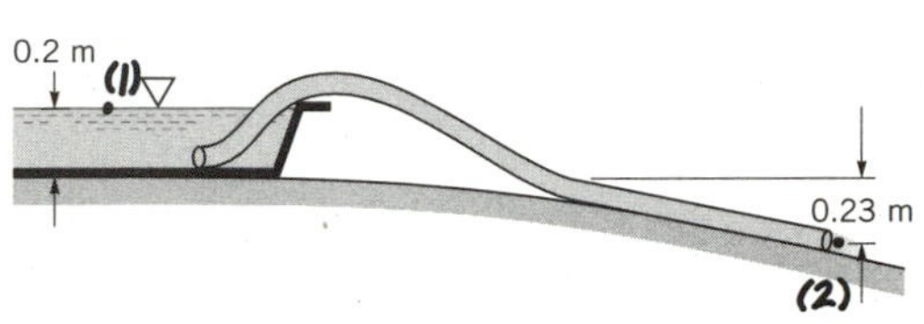

■ FIGURE P3.43

Since $\frac{p_1}{\gamma} + \frac{V_1^2}{2g} + z_1 = \frac{p_2}{\gamma} + \frac{V_2^2}{2g} + z_2$, with $p_1 = p_2 = 0$, $V_1 = 0$, $z_2 = 0$ and $z_1 = 0.2m + 0.23m = 0.43m$, it follows that

$$\frac{V_2^2}{2g} = 0.43\,m$$

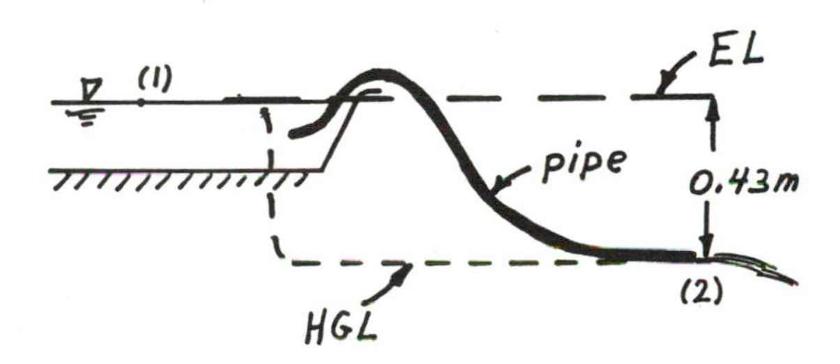

For inviscid flow with no pumps or turbines, the energy line (EL) is horizontal, at an elevation of the free surface. The hydraulic grade line (HGL) is one velocity head lower—even with the pipe outlet. Since the fluid velocity is constant throughout the pipe with $\frac{V^2}{2g} = 0.43m$, the above diagram is obtained.

3.18R

3.18R (Restrictions on Bernoulli equation) A 0.3-m-diameter soccer ball, pressurized to 20 kPa, develops a small leak with an area equivalent to 0.006 mm². If viscous effects are neglected and the air is assumed to be incompressible, determine the flowrate through the hole. Would the ball become noticeably softer during a 1-hr soccer game? Explain. Is it reasonable to assume incompressible flow for this situation? Explain.

(ANS: 9.96×10^{-7} m³/s; yes; no, Ma > 0.3)

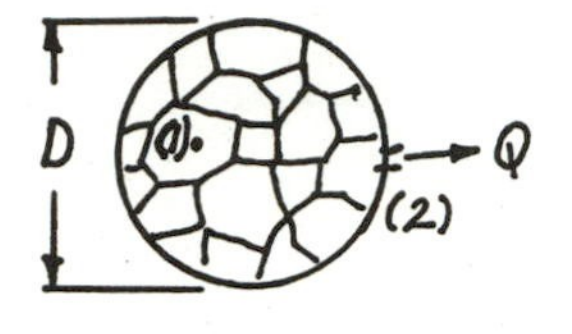

$$\frac{p_1}{\gamma} + \frac{V_1^2}{2g} + z_1 = \frac{p_2}{\gamma} + \frac{V_2^2}{2g} + z_2$$

where $z_1 = z_2$, $p_2 = 0$ and $V_1 = 0$

Thus,

$$\frac{V_2^2}{2g} = \frac{p_1}{\gamma} \quad \text{or} \quad V_2 = \sqrt{\frac{2p_1}{\rho}} \quad \text{where } \rho = \frac{p_1}{RT_1}$$

Assume $T_1 = 15\,°C$ so that

$$\rho = \frac{(101.+20)\times 10^3 \frac{N}{m^2}}{\left(286.9 \frac{N\cdot m}{kg\cdot K}\right)(273+15)K} = 1.46 \frac{kg}{m^3}$$

Thus,

$$V_2 = \sqrt{\frac{2\left(20\times 10^3 \frac{N}{m^3}\right)}{1.46 \frac{kg}{m^3}}} = 166 \frac{m}{s}$$

so that

$$Q = A_2 V_2 = (0.006\ mm^2)\left(\frac{1m}{1000\ mm}\right)^2 \left(166 \frac{m}{s}\right) = \underline{\underline{9.96\times 10^{-7} \frac{m^3}{s}}}$$

If this flowrate continued for one hour, the volume of air leaving the ball would be

$$\forall = Qt = \left(9.96\times 10^{-7} \frac{m^3}{s}\right)(3{,}600\ s) = 3.59\times 10^{-3}\ m^3$$

Since the volume of the ball is

$$\forall_{ball} = \frac{\pi}{6} D^3 = \frac{\pi}{6}(0.3m)^3 = 14.1\times 10^{-3}\ m^3,$$

approximately

$$\frac{\forall}{\forall_{ball}} = \frac{3.59\times 10^{-3} m^3}{14.1\times 10^{-3} m^3} = 0.255 = 25.5\%$$ of the air would escape.

<u>The ball would become noticably softer</u>

Note that $Ma_2 = \frac{V_2}{c_2}$ where $c_2 = \sqrt{kRT_2}$ or if $T_2 \approx 15\,°C$

$$c_2 = \left[1.40\left(286.9 \frac{N\cdot m}{kg\cdot K}\right)(273+15)K\right]^{1/2} = 340 \frac{m}{s}$$

so that the Mach number is

$$Ma_2 = \frac{166 \frac{m}{s}}{340 \frac{m}{s}} = 0.488 > 0.3$$ Thus, should assume <u>compressible flow.</u>

3.19R

3.19R (Restrictions on Bernoulli equation) Niagara Falls is approximately 167 ft high. If the water flows over the crest of the falls with a velocity of 8 ft/s and viscous effects are neglected, with what velocity does the water strike the rocks at the bottom of the falls? What is the maximum pressure of the water on the rocks? Repeat the calculations for the 1430-ft-high Upper Yosemite Falls in Yosemite National Park. Is it reasonable to neglect viscous effects for these falls? Explain.
(ANS: 104 ft/s, 72.8 psi; 304 ft/s, 620 psi; no)

$$\frac{p_1}{\gamma} + \frac{V_1^2}{2g} + z_1 = \frac{p_2}{\gamma} + \frac{V_2^2}{2g} + z_2$$

and

$$\frac{p_1}{\gamma} + \frac{V_1^2}{2g} + z_1 = \frac{p_3}{\gamma} + \frac{V_3^2}{2g} + z_3$$

(1)
h
(2)
(3)

with $p_1 = p_2 = 0$, $V_1 = 8 \frac{ft}{s}$, $V_3 = 0$, $z_2 \approx z_3 = 0$, and $z_1 = h$

Thus,

$$V_2 = \sqrt{2g\left(h + \frac{V_1^2}{2g}\right)} \quad \text{and} \quad p_3 = \frac{1}{2}\rho V_1^2 + \gamma h$$

a) With $h = 167 ft$,

$$V_2 = \left[2\left(32.2\frac{ft}{s^2}\right)\left(167\,ft + \frac{\left(8\frac{ft}{s}\right)^2}{2\left(32.2\frac{ft}{s^2}\right)}\right)\right]^{1/2} = \underline{\underline{104\frac{ft}{s}}}$$

and

$$p_3 = \frac{1}{2}\left(1.94\frac{slugs}{ft^3}\right)\left(8\frac{ft}{s}\right)^2 + 62.4\frac{lb}{ft^3}(167ft) = 10{,}500\frac{lb}{ft^2} = \underline{\underline{72.8\,psi}}$$

b) With $h = 1430 ft$,

$$V_2 = \left[2\left(32.2\frac{ft}{s^2}\right)\left(1430\,ft + \frac{\left(8\frac{ft}{s}\right)^2}{2\left(32.2\frac{ft}{s^2}\right)}\right)\right]^{1/2} = \underline{\underline{304\frac{ft}{s}}}$$

and

$$p_3 = \frac{1}{2}\left(1.94\frac{slugs}{ft^3}\right)\left(8\frac{ft}{s}\right)^2 + 62.4\frac{lb}{ft^3}(1430ft) = 89{,}300\frac{lb}{ft^2} = \underline{\underline{620\,psi}}$$

Aerodynamic drag on the water would reduce the values of V_2 and p_3 (especially for the $h = 1{,}430$ ft case).

4
Fluid Kinematics

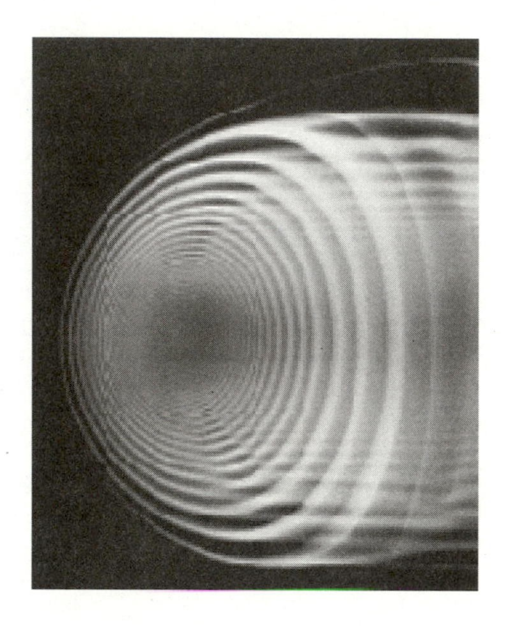

A vortex ring: The complex, three-dimensional structure of a smoke ring is indicated in this cross-sectional view. (Smoke in air.) (Photograph courtesy of R. H. Magarvey and C. S. MacLatchy, Ref. 4.)

4.1R

4.1R (Streamlines) The velocity field in a flow is given by $\mathbf{V} = x^2y\hat{\mathbf{i}} + x^2t\hat{\mathbf{j}}$. **(a)** Plot the streamline through the origin at times $t = 0$, $t = 1$, and $t = 2$. **(b)** Do the streamlines plotted in part (a) coincide with the path of particles through the origin? Explain.

(ANS: $y^2/2 = tx + C$; no)

a) $u = x^2y$ and $v = x^2t$ where the streamlines are obtained from

$\frac{dy}{dx} = \frac{v}{u} = \frac{x^2t}{x^2y}$ or $y\,dy = t\,dx$ which, for a given time t, can be integrated to give

$\frac{1}{2}y^2 = tx + C$, where C is a constant.

For streamlines through the origin $(x=0, y=0)$, $C=0$.

Thus: $y^2 = 2tx$ or

$y = 0$ for $t=0$

$y = \sqrt{2x}$ for $t=1$, and

$y = 2\sqrt{x}$ for $t=2$. These streamlines are plotted below.

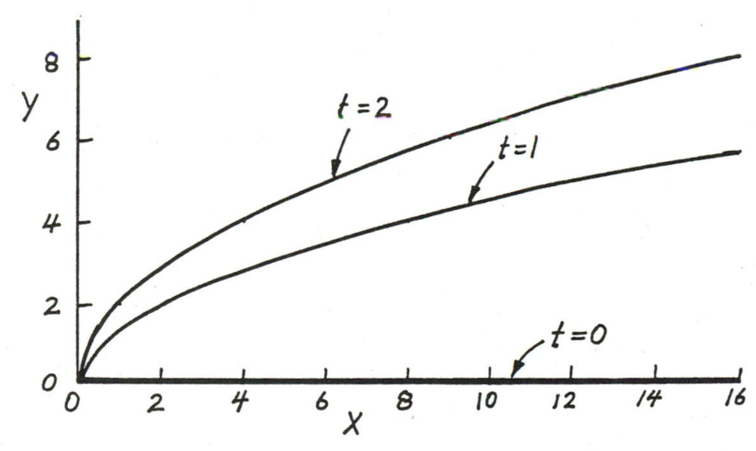

b) Since this is an unsteady flow (i.e. $\frac{\partial v}{\partial t} = x^2 \neq 0$) and since $\frac{dy}{dx} = \frac{t}{y}$ (i.e. at a given location the slope is a function of time), streamlines and pathlines do not coincide.

4.2R

4.2R (Streamlines) A velocity field is given by $u = y - 1$ and $v = y - 2$, where u and v are in m/s and x and y are in meters. Plot the streamline that passes through the point $(x, y) = (4, 3)$. Compare this streamline with the streakline through the point $(x, y) = (4, 3)$.

(ANS: $x = y + \ln(y - 2) + 1$)

$u = y-1$, $v = y-2$ where the streamlines are obtained from

$$\frac{dy}{dx} = \frac{v}{u} = \frac{y-2}{y-1}$$

or

$$\int \frac{(y-1)}{(y-2)}\,dy = \int dx \quad \text{or} \quad \int \frac{y\,dy}{(y-2)} - \int \frac{dy}{(y-2)} = x + \tilde{C}$$

, where $\tilde{C}$ is a constant.

From integral tables:

$$\int \frac{y\,dy}{(y-2)} = y-2+2\ln(y-2) \quad \text{and} \quad \int \frac{dy}{(y-2)} = \ln(y-2)$$

Thus, the streamlines are given by

$$y-2+2\ln(y-2) - \ln(y-2) = x + \tilde{C}$$

or

$$y + \ln(y-2) = x + C \text{ , where } C \text{ is a constant} \qquad (1)$$

For the streamline that passes through $x=4$ and $y=3$, the value of C is found from Eq.(1) as:

$3 + \ln(3-2) = 4 + C$ or $C = -1$ Thus, $\underline{\underline{x = y + \ln(y-2) + 1}}$

This streamline is plotted below:

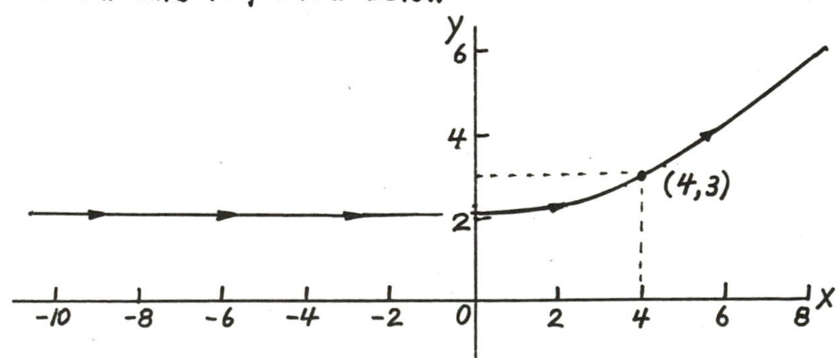

Note: As $x \to -\infty$, $y \to 2$. Also, since $u = y-1$, with $y \geq 2$ anywhere on this streamline, it follows that $u > 0$. The flow is from left to right.

Since the flow is steady, <u>streamlines are the same as streaklines.</u>

4.3R

4.3R (Material derivative) The pressure in the pipe near the discharge of a reciprocating pump fluctuates according to $p = [200 + 40\sin(8t)]$ kPa, where t is in seconds. If the fluid speed in the pipe is 5 m/s, determine the maximum rate of change of pressure experienced by a fluid particle.

(ANS: 320 kPa/s)

Since $u = 5\frac{m}{s}$, $v = 0$, $w = 0$ it follows that with $p = p(t)$

$$\frac{Dp}{Dt} = \frac{\partial p}{\partial t} + u\frac{\partial p}{\partial x} + v\frac{\partial p}{\partial y} + w\frac{\partial p}{\partial z} = \frac{\partial p}{\partial t}$$

or

$$\frac{Dp}{Dt} = 40(8)\cos(8t)\ \frac{kPa}{s} \quad \text{Thus,} \quad \left.\frac{Dp}{Dt}\right)_{max} = 40(8)\ \frac{kPa}{s} = 320\ \frac{kPa}{s}$$

Note: Since $\frac{\partial p}{\partial x} = 0$ the value of u is not important.

4.4R

4.4R (Acceleration) A shock wave is a very thin layer (thickness $= \ell$) in a high-speed (supersonic) gas flow across which the flow properties (velocity, density, pressure, etc.) change from state (1) to state (2) as shown in Fig. P4.4R. If $V_1 = 1800$ fps, $V_2 = 700$ fps, and $\ell = 10^{-4}$ in., estimate the average deceleration of the gas as it flows across the shock wave. How many g's deceleration does this represent?

(ANS: -1.65×10^{11} ft/s^2; -5.12×10^{9})

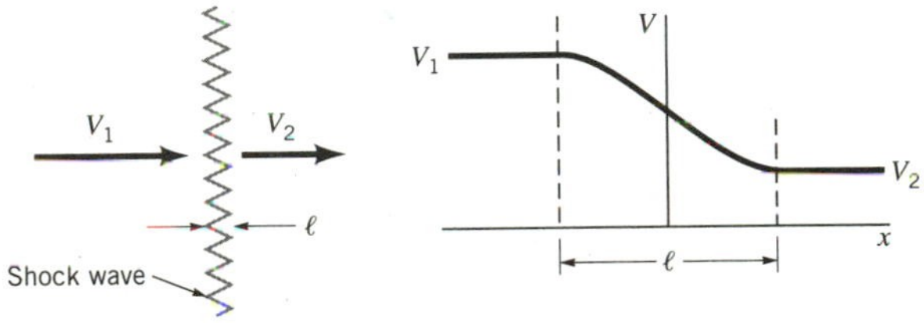

■ FIGURE P4.4R

$$\vec{a} = \frac{\partial \vec{V}}{\partial t} + \vec{V}\cdot\nabla\vec{V} \quad \text{so with} \quad \vec{V} = u(x)\hat{i}, \quad \vec{a} = a_x\hat{i} = u\frac{\partial u}{\partial x}\hat{i}$$

Without knowing the actual velocity distribution, $u = u(x)$, the acceleration can be approximated as

$$a_x = u\frac{\partial u}{\partial x} \approx \frac{(V_1 + V_2)}{2}\ \frac{(V_2 - V_1)}{\ell} = \frac{(1800 + 700)\,fps}{2}\ \frac{(700 - 1800)\,fps}{\left(\frac{10^{-4}}{12}\right)ft}$$

or

$$a_x = -1.65 \times 10^{11}\ \frac{ft}{s^2} \qquad \text{This is} \quad \frac{a_x}{g} = \frac{-1.65 \times 10^{11}\ \frac{ft}{s^2}}{32.2\ \frac{ft}{s^2}} = -5.12 \times 10^{9}$$

4.5R

4.5R (Acceleration) Air flows through a pipe with a uniform velocity of $\mathbf{V} = 5\,t^2\hat{\mathbf{i}}$ ft/s, where t is in seconds. Determine the acceleration at time $t = -1$, 0, and 1 s.

(ANS: $-10\,\hat{\mathbf{i}}$ ft/s²; 0; $10\,\hat{\mathbf{i}}$ ft/s²)

$\vec{a} = \frac{\partial \vec{V}}{\partial t} + \vec{V}\cdot\nabla\vec{V}$ With $u = 5t^2\,\frac{ft}{s}$, $v=0$, $w=0$

this becomes

$$\vec{a} = \left(\frac{\partial u}{\partial t} + u\frac{\partial u}{\partial x}\right)\hat{\imath} = \frac{\partial u}{\partial t}\hat{\imath} = 10t\,\hat{\imath}\,\frac{ft}{s^2} \quad \text{since } \frac{\partial u}{\partial x} = 0.$$

Thus, $\vec{a} = \underline{\underline{-10\hat{\imath}\,\frac{ft}{s^2}}}$ at $t = -1$ s

$\vec{a} = \underline{\underline{0}}$ at $t = 0$

and

$\vec{a} = \underline{\underline{10\hat{\imath}\,\frac{ft}{s^2}}}$ at $t = 1$ s

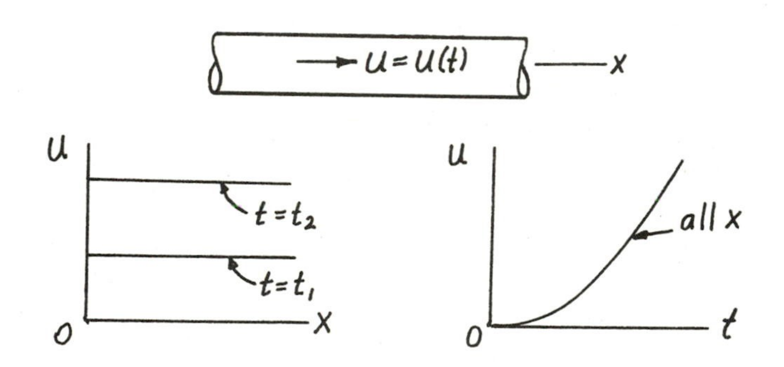

4.6R

4.6R (Acceleration) A fluid flows steadily along the streamline as shown in Fig. P4.6R. Determine the acceleration at point A. At point A what is the angle between the acceleration and the x axis? At point A what is the angle between the acceleration and the streamline?

(ANS: $10\,\hat{\mathbf{n}} + 30\,\hat{\mathbf{s}}$ ft/s²; 48.5 deg; 18.5 deg)

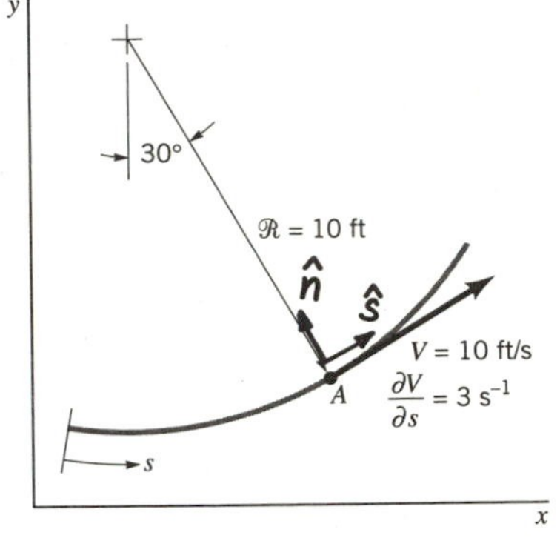

■ FIGURE P4.6R

$$\vec{a} = a_n \hat{n} + a_s \hat{s} = \frac{V^2}{\mathcal{R}}\hat{n} + V\frac{\partial V}{\partial s}\hat{s} = \frac{(10\frac{ft}{s})^2}{10ft}\hat{n} + (10\tfrac{ft}{s})(3\tfrac{1}{s})\hat{s}$$

or

$$\vec{a} = 10\hat{n} + 30\hat{s}\ \frac{ft}{s^2}$$

In terms of unit vectors $\hat{i}$ and $\hat{j}$, $\hat{n} = -\sin 30^\circ \hat{i} + \cos 30^\circ \hat{j}$

and $\hat{s} = \cos 30^\circ \hat{i} + \sin 30^\circ \hat{j}$

Thus,

$$\vec{a} = 10(-0.5\hat{i} + 0.866\hat{j}) + 30(0.866\hat{i} + 0.5\hat{j}) = 21.0\hat{i} + 23.7\hat{j}\ \frac{ft}{s^2}$$

Hence, $\theta = \tan^{-1}\frac{a_y}{a_x} = \tan^{-1}\frac{23.7}{21.0}$

or $\theta = 48.5^\circ$

and

$\alpha = \theta - 30^\circ = 18.5^\circ$

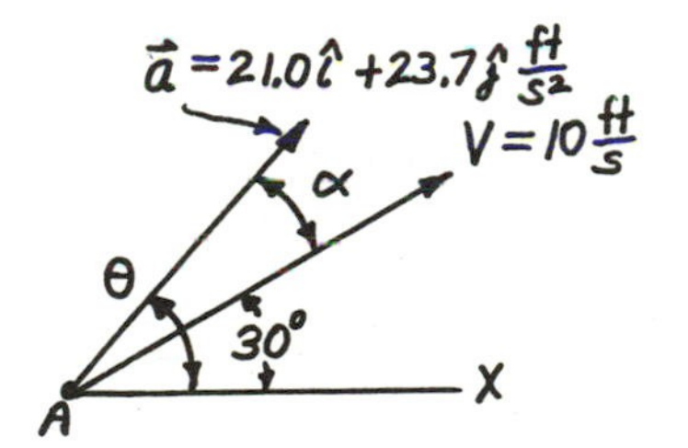

4.7R

4.7R (Acceleration) In the conical nozzle shown in Fig. P4.7R the streamlines are essentially radial lines emanating from point A and the fluid velocity is given approximately by $V = C/r^2$, where C is a constant. The fluid velocity is 2 m/s along the centerline at the beginning of the nozzle ($x = 0$). Determine the acceleration along the nozzle centerline as a function of x. What is the value of the acceleration at $x = 0$ and $x = 0.3$ m?

(ANS: $1.037/(0.6 - x)^5\ \hat{i}$ m/s²; $13.3\ \hat{i}$ m/s²; $427\ \hat{i}$ m/s²)

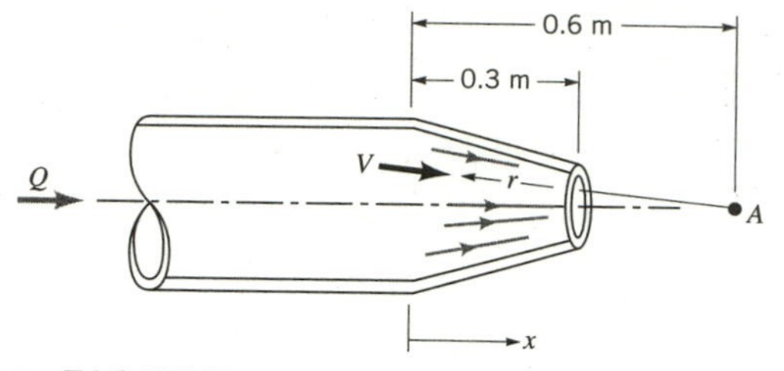

■ FIGURE P4.7R

Along the nozzle centerline, $\vec{a} = \frac{\partial \vec{V}}{\partial t} + \vec{V}\cdot\nabla\vec{V}$ becomes

$\vec{a} = u\frac{\partial u}{\partial x}\hat{i}$ where $u = \frac{C}{r^2}$ with $r + x = 0.6\,m$, or $r = 0.6 - x$.

Thus,

$$u = \frac{C}{(0.6-x)^2}$$

Since $u = 2\frac{m}{s}$ at $x = 0$ it follows that

$$2\frac{m}{s} = \frac{C}{(0.6m)^2} \quad \text{or} \quad C = 0.72\frac{m^3}{s}$$

Hence,

$$\vec{a} = \left[\frac{C}{(0.6-x)^2}\right]\frac{2C}{(0.6-x)^3}\hat{i} = \frac{2C^2}{(0.6-x)^5}\hat{i} = \frac{2(0.72)^2\frac{m^6}{s^2}}{(0.6-x)^5 m^5}\hat{i}$$

or

$$\vec{a} = \frac{1.037}{(0.6-x)^5}\hat{i}\frac{m}{s^2} \quad \text{where } x \sim m$$

At $x = 0$, $\vec{a} = 13.3\,\hat{i}\,\frac{m}{s^2}$; at $x = 0.3\,m$, $\vec{a} = 427\,\hat{i}\,\frac{m}{s^2}$

4.8R

4.8R (Reynolds transport theorem) A sanding operation injects 10^5 particles/s into the air in a room as shown in Fig. P4.8R. The amount of dust in the room is maintained at a constant level by a ventilating fan that draws clean air into the room at section (1) and expels dusty air at section (2). Consider a control volume whose surface is the interior surface of the room (excluding the sander) and a system consisting of the material within the control volume at time $t = 0$. **(a)** If N is the number of particles, discuss the physical meaning of and evaluate the terms DN_{sys}/Dt and $\partial N_{cv}/\partial t$. **(b)** Use the Reynolds transport theorem to determine the concentration of particles (particles/m³) in the exhaust air for steady state conditions.

(ANS: 5×10^5 particles/m³)

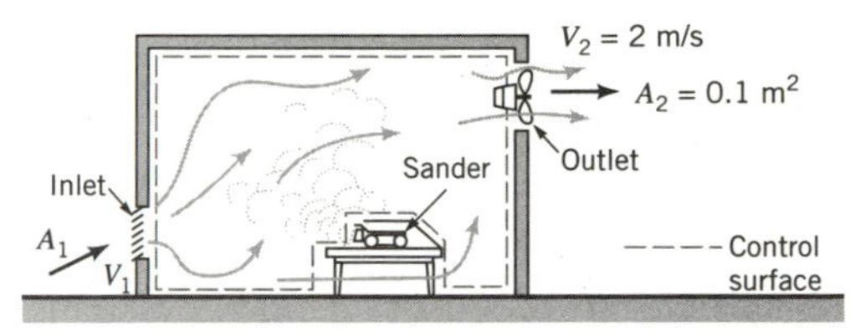

■ FIGURE P4.8R

a) $\frac{DN_{sys}}{Dt}$ = time rate of change of the number of particles in the system. At time $t=0$ the system consists of N particles. In fact, the system is these particles for all time $t>0$. Assuming that the particles do not get "glued" together, N remains constant.

Thus, $\frac{DN_{sys}}{Dt} = 0$

$\frac{\partial N_{cv}}{\partial t}$ = time rate of change of the number of particles in the control volume. Depending on the rate at which the sander creates particles and ejects them into the room compared to the rate at which the fan draws them from the room, we could have $\frac{\partial N_{cv}}{\partial t} \gtreqless 0$.

b) $\frac{DN_{sys}}{Dt} = \frac{\partial N_{cv}}{\partial t}$ + net rate of flow of particles out of control volume

or for steady state $\frac{\partial N_{cv}}{\partial t} = 0$ so that:

flow of particles into control volume (from sander, none enter A_1)
= flow of particles out of control volume (through fan exhaust, A_2)

Thus,

$$10^5 \frac{\text{particles}}{s} = V_2 A_2 n_2$$, where n_2 = particle concentration $\left(\frac{\text{particles}}{m^3}\right)$

Hence,

$$n_2 = \frac{10^5 \frac{\text{particles}}{s}}{(2 \frac{m}{s})(0.1 m^2)} = 5 \times 10^5 \frac{\text{particles}}{m^3}$$

4.9R (Flowrate) Water flows through the rectangular channel shown in Fig. P4.9R with a uniform velocity. Directly integrate Eqs. 4.16 and 4.17 with $b = 1$ to determine the mass flowrate (kg/s) across and A–B of the control volume. Repeat for C–D.

(ANS: 18,000 kg/s; 18,000 kg/s)

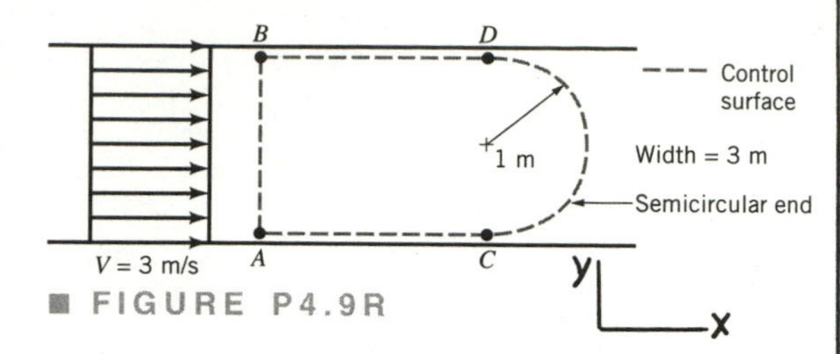

■ FIGURE P4.9R

Equation 4.17:

$$\dot{B}_{in} = -\int_{AB} \rho b \vec{V}\cdot\hat{n}\, dA \quad \text{or, with } \rho = 999\,\frac{kg}{m^3} \text{ and } b=1 \text{ this gives}$$

$$\dot{B}_{in} = -999\,\frac{kg}{m^3}\int (3\,\tfrac{m}{s})\hat{i}\cdot(\hat{i})\,dA = -(999\,\tfrac{kg}{m^3})(3\,\tfrac{m}{s})(2m)(3m)$$

$$= 18{,}000\,\frac{kg}{s}$$

and Equation 4.16:

$$\dot{B}_{out} = \int_{CD} \rho b \vec{V}\cdot\hat{n}\, dA = \rho\int_{CD} \vec{V}\cdot\hat{n}\, dA$$

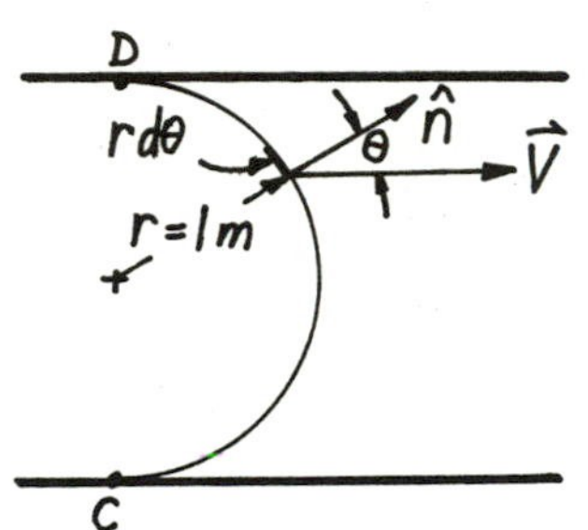

where $dA = 3\, r d\theta = 3\, d\theta$

and $\vec{V}\cdot\hat{n} = V\cos\theta = 3\cos\theta$

Thus,

$$\dot{B}_{out} = 999\,\frac{kg}{m^3}\int_{\theta=-90^\circ}^{\theta=90^\circ} (3\cos\theta\,\tfrac{m}{s})(3\,m)(1\,m)\,d\theta = 9(999)\sin\theta\Big|_{-90^\circ}^{90^\circ}$$

or

$$\dot{B}_{out} = 18{,}000\,\frac{kg}{s}$$

4.10R

4.10R (Flowrate) Air blows through two windows as indicated in Fig. P4.10R. Use Eq. 4.16 with $b = 1/\rho$ to determine the volume flowrate (ft^3/s) through each window. Explain the relationship between the two results you obtained.

(ANS: 80 ft^3/s; 160 ft^3/s)

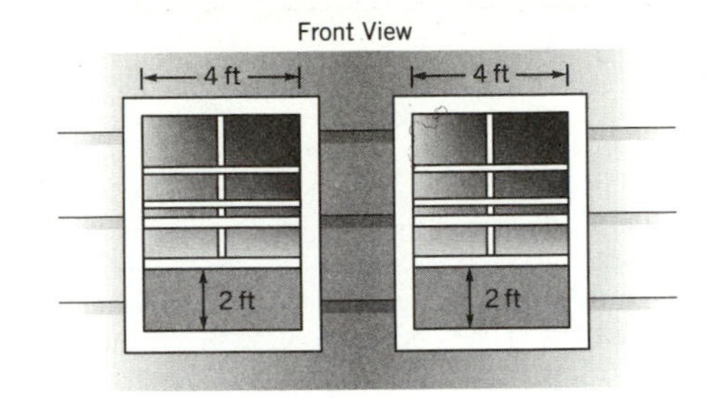

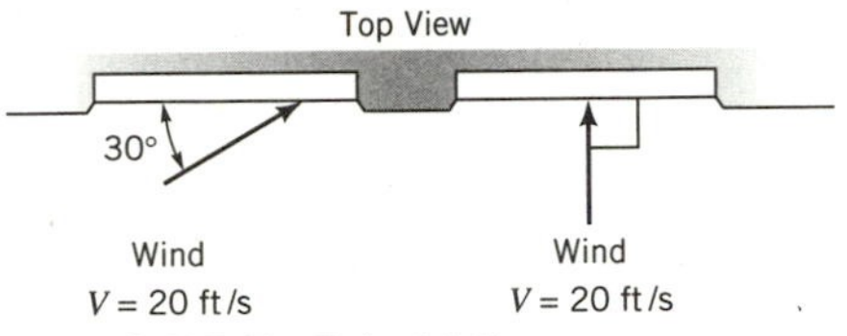

Wind
$V = 20$ ft/s

Wind
$V = 20$ ft/s

■ FIGURE P4.10R

$$\dot{B} = \int \rho b \vec{V} \cdot \hat{n} \, dA$$

For the left window $Q = \int \rho \left(\frac{1}{\rho}\right) \vec{V} \cdot \hat{n} \, dA = \int \vec{V} \cdot \hat{n} \, dA$

or $Q = \int 20 \sin 30^\circ \, dA = \left(20 \sin 30^\circ \frac{ft}{s}\right)(4\,ft)(2\,ft) = \underline{\underline{80 \frac{ft^3}{s}}}$

For the right window $Q = \int \vec{V} \cdot \hat{n} \, dA$

or $Q = \int 20 \, dA = \left(20 \frac{ft}{s}\right)(4\,ft)(2\,ft) = \underline{\underline{160 \frac{ft^3}{s}}}$

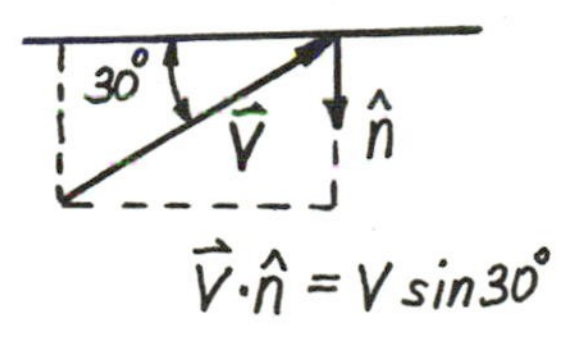

$\vec{V} \cdot \hat{n} = V \sin 30^\circ$

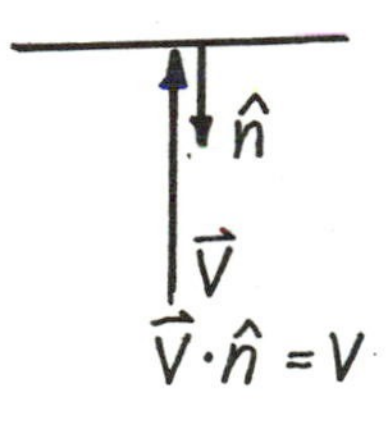

$\vec{V} \cdot \hat{n} = V$

4.11R

4.11R (Control volume/system) Air flows over a flat plate with a velocity profile given by $\mathbf{V} = u(y)\hat{\mathbf{i}}$, where $u = 2y$ ft/s for $0 \leq y \leq 0.5$ ft and $u = 1$ ft/s for $y > 0.5$ ft as shown in Fig. P4.11R. The fixed rectangular control volume $ABCD$ coincides with the system at time $t = 0$. Make a sketch to indicate **(a)** the boundary of the system at time $t = 0.1$ s, **(b)** the fluid that moved out of the control volume in the interval $0 \leq t \leq 0.1$ s, and **(c)** the fluid that moved into the control volume during that time interval.

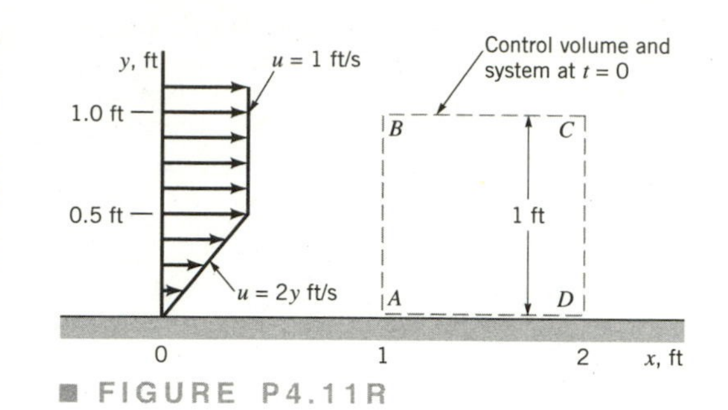

■ FIGURE P4.11R

Since $\vec{V} = u(y)\hat{i}$, each fluid particle travels only in the x-direction, with the distance of travel $\delta x = u\,\delta t$, where $\delta t = 0.1$ s.
Thus, $\delta x_A = \delta x_D = 0$ since $u = 0$ at $y = 0$.
Also $\delta x_B = \delta x_C = (1\,\frac{ft}{s})(0.1\,s) = 0.1$ ft.
The fluid originally along lines A-B and C-D move to positions A-E-B′ and D-F-C′ as shown it the figure below. The location of the system at $t = 0.1$ s and the fluid that moved into or out of the control volume is indicated.

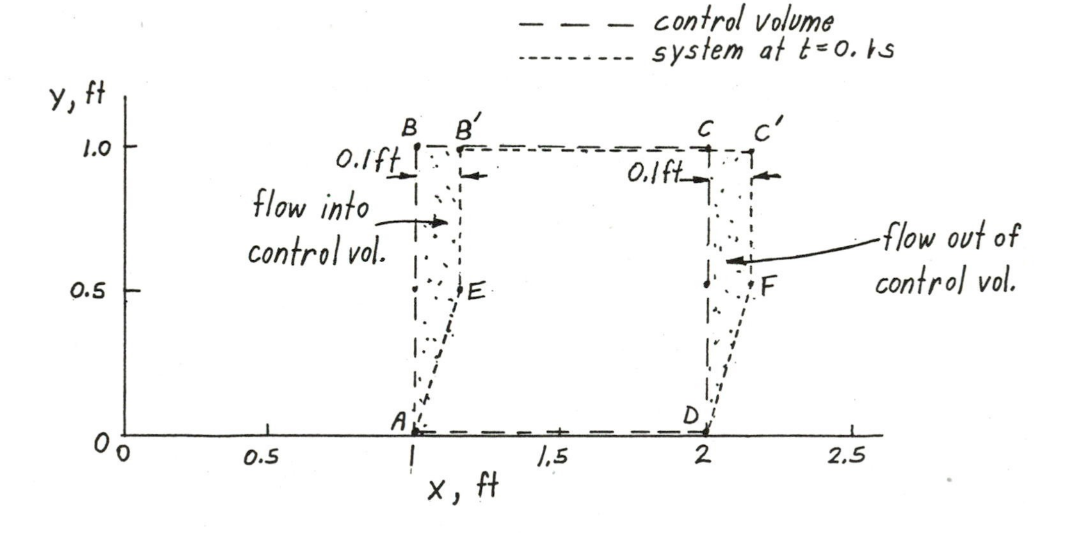

5
Finite Control Volume Analysis

A jet of water injected into stationary water: Upon emerging from the slit at the left, the jet of fluid loses some of its momentum to the surrounding fluid. This causes the jet to slow down and its width to increase (air bubbles in water). (Photograph courtesy of ONERA, France.)

5.1R (Continuity equation) Water flows steadily through a 2-in.-inside-diameter pipe at the rate of 200 gal/min. The 2-in. pipe branches into two 1-in.-inside-diameter pipes. If the average velocity in one of the 1-in. pipes is 30 ft/s, what is the average velocity in the other 1-in. pipe?
(ANS: 51.7 ft/s)

(1)
2 in.
$Q_1 = 200\ gpm$
(2)
1 in.
$\overline{V}_2$
1 in.
(3)
$\overline{V}_3 = 30\ \frac{ft}{s}$

For steady incompressible flow

$$Q_1 = Q_2 + Q_3$$

or

$$Q_1 = A_2\overline{V}_2 + A_3\overline{V}_3$$

Thus

$$\overline{V}_2 = \frac{Q_1}{A_2} - \frac{A_3\,\overline{V}_3}{A_2} = \frac{Q_1}{\frac{\pi D_2^2}{4}} - \frac{D_3^2}{D_2^2}\overline{V}_3$$

$$\overline{V}_2 = \frac{(200\ gpm)\left(231\ \frac{in.^3}{gal}\right)}{\frac{\pi}{4}(1\ in.)^2\left(60\ \frac{s}{min}\right)\left(12\ \frac{in.}{ft}\right)} - \frac{(1\ in.)^2\left(30\ \frac{ft}{s}\right)}{(1\ in.)^2} = \underline{\underline{51.7\ \frac{ft}{s}}}$$

5.2R

5.2R (Continuity equation) Air (assumed incompressible) flows steadily into the square inlet of an air scoop with the nonuniform velocity profile indicated in Fig. P5.2R. The air exits as a uniform flow through a round pipe 1 ft in diameter. **(a)** Determine the average velocity at the exit plane. **(b)** In one minute, how many pounds of air pass through the scoop?

(ANS: 191 ft/s; 688 lb/min)

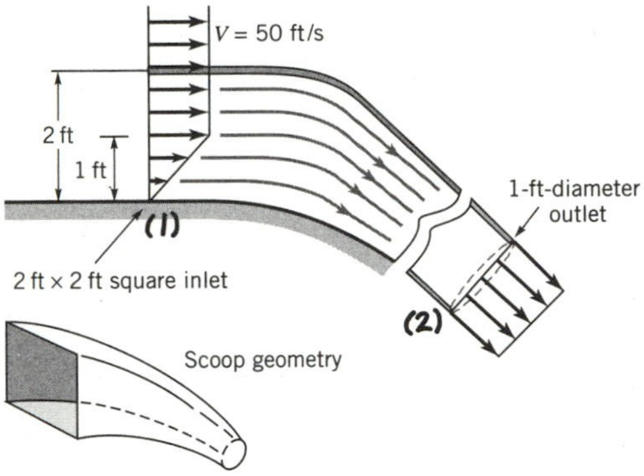

FIGURE P5.2R

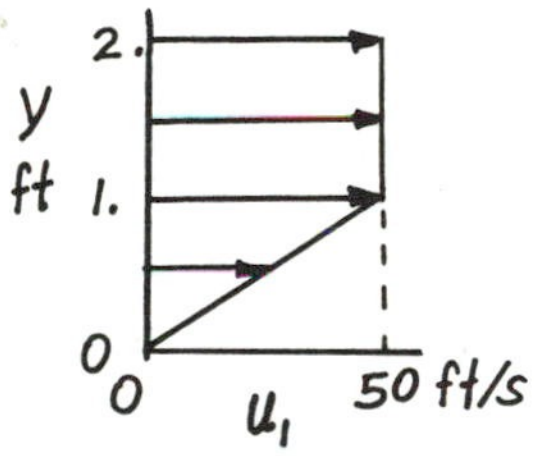

a) For steady, incompressible flow

$Q_1 = Q_2$ or

$$\int_{(1)} u_1\,dA = A_2 V_2 \quad \text{where} \quad \int_{(1)} u_1\,dA = \int_{y=0}^{y=1.} u_1 (2\,dy) + \int_{y=1.}^{y=2.} (50)(2\,dy)$$

$$= (2\,ft)(1\,ft)\left(\frac{50}{2}\frac{ft}{s}\right) + (2\,ft)(1\,ft)\left(50\frac{ft}{s}\right)$$

$$= 150\frac{ft^3}{s} = Q$$

Thus,

$$V_2 = \frac{150\frac{ft^3}{s}}{\frac{\pi}{4}(1\,ft)^2} = \underline{\underline{191\frac{ft}{s}}}$$

b) Weight flow rate $= \gamma Q = \left(0.0765\frac{lb}{ft^3}\right)\left(150\frac{ft^3}{s}\right) = 11.48\frac{lb}{s}\left(\frac{60\,s}{min}\right) = \underline{\underline{688\frac{lb}{min}}}$

Note: $$\int_{y=0}^{y=1.} u_1 (2\,dy) = \int_0^1 (50y)(2\,dy) = 100\int_0^1 y\,dy = 50y^2\Big|_0^1 = 50\frac{ft^3}{s}$$

5.3R

5.3R (Continuity equation) Water at 0.1 m^3/s and alcohol (SG = 0.8) at 0.3 m^3/s are mixed in a y-duct as shown in Fig. P5.3R. What is the average density of the mixture of alcohol and water?

(ANS: 849 kg/m^3)

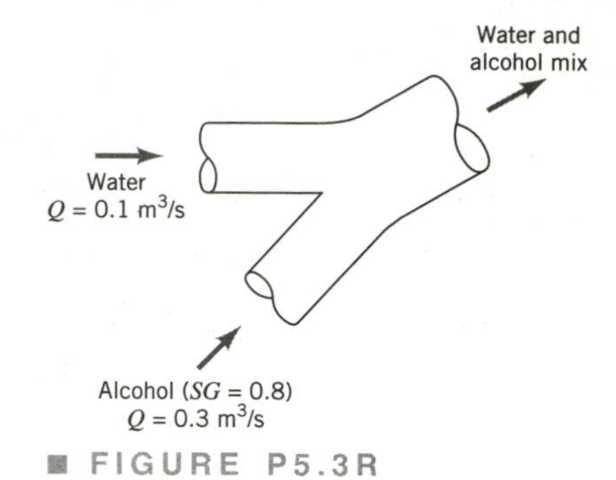

■ FIGURE P5.3R

Q_1, ρ_1 (1); ρ_2, Q_2 (2); ρ_3, Q_3 (3); control surface

For steady flow $\int_{cs} \rho \vec{V} \cdot \hat{n} \, dA = 0$, or

$$\dot{m}_1 + \dot{m}_2 = \dot{m}_3$$

or

$$\rho_1 Q_1 + \rho_2 Q_2 = \rho_3 Q_3 \qquad (1)$$

Also, since the water and alcohol may be considered incompressible

$$Q_1 + Q_2 = Q_3 \qquad (2)$$

Combining Eqs. 1 and 2 we get

$$\rho_1 Q_1 + \rho_2 Q_2 = \rho_3 (Q_1 + Q_2)$$

or

$$\rho_3 = \frac{\rho_1 Q_1 + \rho_2 Q_2}{Q_1 + Q_2}$$

and

$$\rho_3 = \rho_1 \frac{(Q_1 + SG_2 Q_2)}{Q_1 + Q_2}$$

Thus

$$\rho_3 = \frac{\left(999 \frac{kg}{m^3}\right)\left[0.1 \frac{m^3}{s} + (0.8)\left(0.3 \frac{m^3}{s}\right)\right]}{0.1 \frac{m^3}{s} + 0.3 \frac{m^3}{s}} = \underline{\underline{849 \frac{kg}{m^3}}}$$

5.4R (Average velocity) The flow in an open channel has a velocity distribution

$$\mathbf{V} = U(y/h)^{1/5}\hat{\mathbf{i}} \text{ ft/s}$$

where U = free-surface velocity, y = perpendicular distance from the channel bottom in feet, and h = depth of the channel in feet. Determine the average velocity of the channel stream as a fraction of U.

(ANS: 0.833)

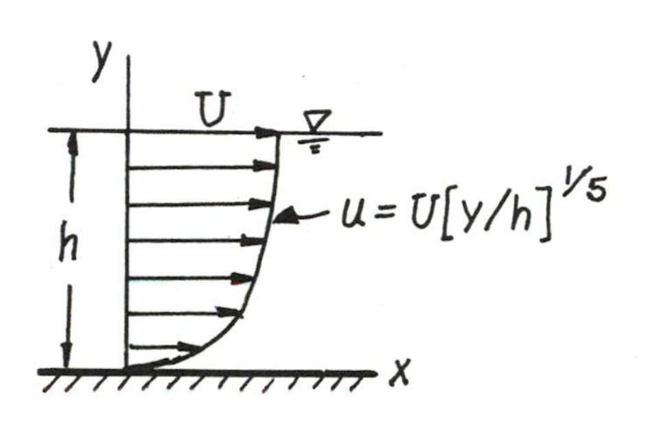

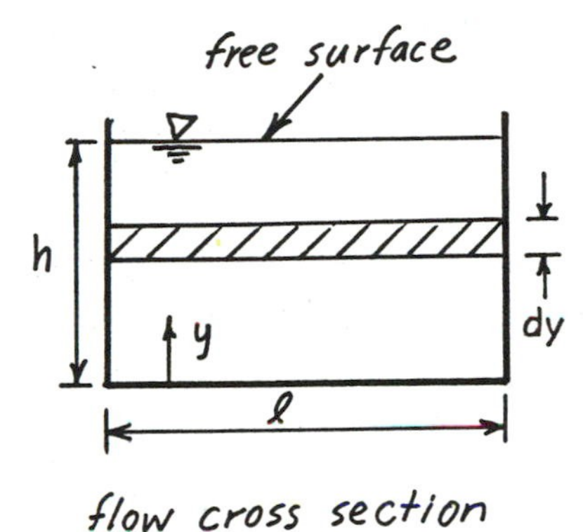

flow cross section

For any flow cross section

$$\dot{m} = \rho A \bar{u} = \int_A \rho \vec{V} \cdot \hat{n} \, dA$$

Also

$$\vec{V} \cdot \hat{n} = \vec{V} \cdot \hat{i} = U\left(\frac{y}{h}\right)^{\frac{1}{5}}$$

Thus for uniformly distributed density, ρ, over area A

$$\bar{u} = \frac{\int_0^h U\left(\frac{y}{h}\right)^{\frac{1}{5}} \ell \, dy}{\ell h}$$

and

$$\frac{\bar{u}}{U} = \int_0^1 \left(\frac{y}{h}\right)^{\frac{1}{5}} d\left(\frac{y}{h}\right) = \frac{5}{6} = \underline{\underline{0.833}}$$

5.5R

5.5R (Linear momentum) Water flows through a right angle valve at the rate of 1000 lbm/s as is shown in Fig. P5.5R. The pressure just upstream of the valve is 90 psi and the pressure drop across the valve is 50 psi. The inside diameters of the valve inlet and exit pipes are 12 and 24 in. If the flow through the valve occurs in a horizontal plane, determine the x and y components of the force exerted by the valve on the water.
(ANS: 18,200 lb; 10,800 lb)

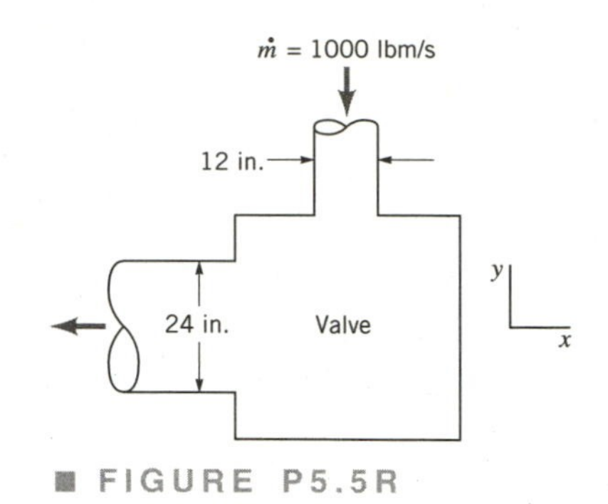

■ FIGURE P5.5R

For steady flow the x-component of the momentum equation is

$$\int_{cs} u\rho\vec{V}\cdot\hat{n}\,dA = \Sigma F_x$$

or

$$u_1\rho(u_1\hat{i}+v_1\hat{j})\cdot\hat{j}A_1 + u_2\rho(u_2\hat{i}+v_2\hat{j})\cdot(-\hat{i})A_2 = +p_2A_2 - F_{Ax}$$

but $u_1=0$, $v_1=-V_1$, $u_2=-V_2$, and $v_2=0$

Thus,

$$0+(-V_2)\rho(V_2)A_2 = +p_2A_2 - F_{Ax}$$

or

$$F_{Ax} = p_2A_2 + \dot{m}V_2, \text{ where } \dot{m} = \rho V_2A_2 = \rho V_1A_1 = 30\,\frac{\text{slugs}}{\text{s}}$$

$$\vec{V} = u\hat{i} + v\hat{j}$$

Hence,

$$F_{Ax} = (90-50)\frac{\text{lb}}{\text{in.}^2}\left(\frac{\pi}{4}(24\text{ in.})^2\right) + 30\,\frac{\text{slugs}}{\text{s}}\left(4.92\,\frac{\text{ft}}{\text{s}}\right) = \underline{\underline{18{,}200\text{ lb}}}$$

where we have used $V_2 = \dfrac{\dot{m}}{A_2\rho}$

or

$$V_2 = \frac{30\,\frac{\text{slugs}}{\text{s}}}{\left(\frac{\pi}{4}(2\text{ ft})^2\right)\left(1.94\,\frac{\text{slugs}}{\text{ft}^3}\right)} = 4.92\,\frac{\text{ft}}{\text{s}}$$

(continued)

5.5R continued

Similarly, in the y-direction $\int_{cs} v \rho \vec{V}\cdot\hat{n}\,dA = \Sigma F_y$, or

$$v_1 \rho(u_1\hat{i}+v_1\hat{j})\cdot\hat{j}\,A_1 + v_2\rho(u_2\hat{i}+v_2\hat{j})\cdot(-\hat{i})A_2 = -p_1A_1 + F_{Ay}$$

or

$$(-V_1)\rho(-V_1)A_1 + 0 = -p_1A_1 + F_{Ay}\text{, or } F_{Ay} = p_1A_1 + \rho A_1V_1^2 = p_1A_1 + \dot{m}V_1$$

Thus, since $V_1 = \frac{A_2V_2}{A_1} = \left(\frac{D_2}{D_1}\right)^2 V_2 = \left(\frac{24\text{ in.}}{12\text{ in.}}\right)^2\left(4.92\ \frac{\text{ft}}{\text{s}}\right) = 19.7\ \frac{\text{ft}}{\text{s}}$ we obtain

$$F_{Ay} = 90\ \frac{\text{lb}}{\text{in.}^2}\left(\frac{\pi}{4}(12\text{ in.})^2\right) + 30\ \frac{\text{slugs}}{\text{s}}\left(19.7\ \frac{\text{ft}}{\text{s}}\right) = \underline{\underline{10{,}800\text{ lb}}}$$

5.6R

5.6R (Linear momentum) A horizontal circular jet of air strikes a stationary flat plate as indicated in Fig. P5.6R. The jet velocity is 40 m/s and the jet diameter is 30 mm. If the air velocity magnitude remains constant as the air flows over the plate surface in the directions shown, determine: **(a)** the magnitude of F_A, the anchoring force required to hold the plate stationary, **(b)** the fraction of mass flow along the plate surface in each of the two directions shown, **(c)** the magnitude of F_A, the anchoring force required to allow the plate to move to the right at a constant speed of 10 m/s.

(ANS: 0.696 N; 0.933 and 0.0670; 0.391 N)

$D_j = 30$ mm, $V_j = 40$ m/s, V_2, V_3, 30°, 90°, F_A

■ FIGURE P5.6R

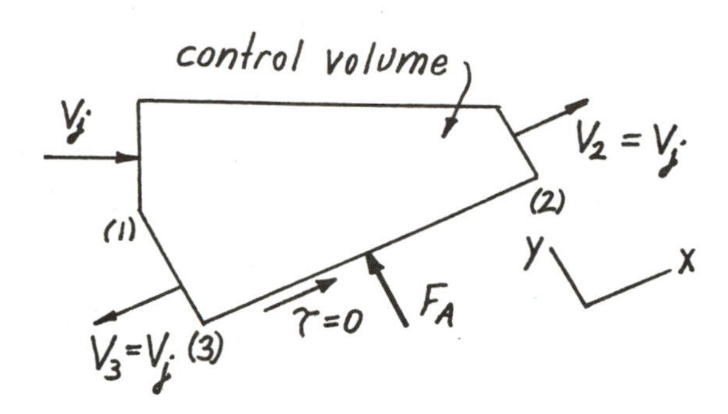

The non-deforming control volume shown in the sketch above is used.

(a) To determine the magnitude of F_A we apply the component of the linear momentum equation (Eq. 5.22) along the direction of F_A. Thus, $\int_{cs} v \rho \vec{V} \cdot \hat{n} \, dA = \Sigma F_y$, or

$$F_A = \dot{m} V_j \sin 30° = \rho A_j V_j V_j \sin 30° = \frac{\rho \pi D_j^2}{4} V_j^2 \sin 30°$$

or

$$F_A = \left(1.23 \frac{kg}{m^3}\right) \frac{\pi (0.030 m)^2}{(4)} \left(40 \frac{m}{s}\right)^2 (\sin 30°) \left(\frac{1 \, N}{kg \cdot \frac{m}{s^2}}\right) = \underline{\underline{0.696 \, N}}$$

(b) To determine the fraction of mass flow along the plate surface in each of the 2 directions shown in the sketch above, we apply the component of the linear momentum equation parallel to the surface of the plate, $\int_{cs} u \rho \vec{V} \cdot \hat{n} \, dA = \Sigma F_x$, to obtain

$$R_{\text{along plate surface}} = \dot{m}_2 V_2 - \dot{m}_3 V_3 - \dot{m}_j V_j \cos 30° \qquad (1)$$

(continued)

5.6R continued

Since the air velocity magnitude remains constant, the value of $R_{\substack{along\ plate\\surface}}$ is zero.* Thus from Eq. 1 we obtain

$$\dot{m}_3 V_3 = \dot{m}_2 V_2 - \dot{m}_j V_j \cos 30° \qquad (2)$$

Since $V_3 = V_2 = V_j$, Eq. 2 becomes

$$\dot{m}_3 = \dot{m}_2 - \dot{m}_j \cos 30° \qquad (3)$$

From conservation of mass we conclude that

$$\dot{m}_j = \dot{m}_2 + \dot{m}_3 \qquad (4)$$

Combining Eqs. 3 and 4 we get

$$\dot{m}_3 = \dot{m}_j - \dot{m}_3 - \dot{m}_j \cos 30°$$

or

$$\dot{m}_3 = \dot{m}_j \frac{(1 - \cos 30°)}{2} = \dot{m}_j (0.0670)$$

and

$$\dot{m}_2 = \dot{m}_j (1 - 0.067) = \dot{m}_j (0.933)$$

Thus, $\underline{\underline{\dot{m}_2 \text{ involves } 93.3\% \text{ of } \dot{m}_j}}$ and $\underline{\underline{\dot{m}_3 \text{ involves } 6.7\% \text{ of } \dot{m}_j}}$.

(c) To determine the magnitude of F_A required to allow the plate to move to the right at a constant speed of $10 \frac{m}{s}$, we use a non-deforming control volume like the one in the sketch above that moves to the right with a speed of $10 \frac{m}{s}$. The translating control volume linear momentum equation (Eq. 5.29) leads to

$$F_A = \frac{\rho \pi D_j^2}{4} \left(V_j - 10 \frac{m}{s}\right)^2 \sin 30°$$

or

$$F_A = \left(1.23 \frac{kg}{m^3}\right) \frac{\pi (0.030\, m)^2}{4} \left(40 \frac{m}{s} - 10 \frac{m}{s}\right)^2 (\sin 30°) \left(1 \frac{N}{kg \cdot \frac{m}{s^2}}\right)$$

and

$$F_A = \underline{\underline{0.391\ N}}$$

---

* Since $V_1 = V_2 = V_3$ and $p_1 = p_2 = p_3$ and $z_1 = z_2 = z_3$ it follows that the Bernoulli equation is valid from $1 \rightarrow 2$ and $1 \rightarrow 3$. Thus, there are no viscous effects (Bernoulli equation is valid only for inviscid flow) so that $\tau = 0$. Hence, $R_{along\ plate} = 0$.

5.7R

5.7R (Linear momentum) An axisymmetric device is used to partially "plug" the end of the round pipe shown in Fig. P5.7R. The air leaves in a radial direction with a speed of 50 ft/s as indicated. Gravity and viscous forces are negligible. Determine the **(a)** flowrate through the pipe, **(b)** gage pressure at point (1), **(c)** gage pressure at the tip of the plug, point (2), **(d)** force, F, needed to hold the plug in place.
(ANS: 23.6 ft³/s; 1.90 lb/ft²; 2.97 lb/ft²; 3.18 lb)

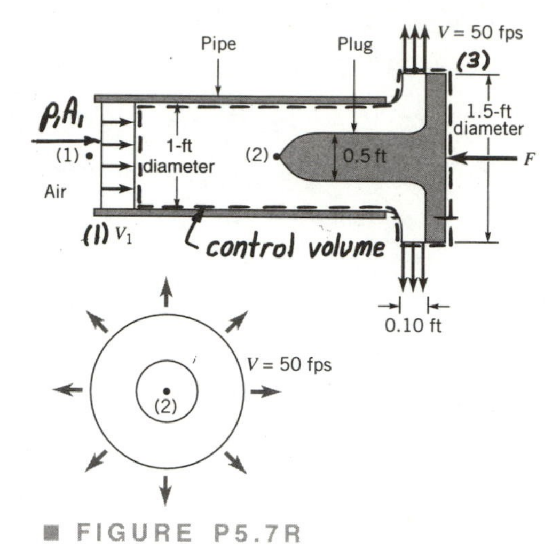

■ FIGURE P5.7R

For part (a) we determine the volume flowrate through the pipe by calculating the volume flowrate of the air leaving radially after being turned by the axisymmetric plug. Thus

$$Q = V_3 A_3 = \left(50\ \frac{ft}{s}\right)\pi(1.5\ ft)(0.10\ ft) = \underline{\underline{23.6\ \frac{ft^3}{s}}}$$

For part (b) we determine the gage pressure at (1) by applying the Bernoulli equation to the flow between (1) and the radial flow leaving the plug, station (3). Thus

$$\frac{p_1}{\rho} + \frac{V_1^2}{2} = \frac{p_3}{\rho} + \frac{V_3^2}{2} \quad (p_3 = 0\ \text{gage}) \tag{1}$$

We get V_1 from

$$V_1 = \frac{Q}{A_1} = \frac{23.6\ \frac{ft^3}{s}}{\frac{\pi(1\ ft)^2}{4}} = 30\ \frac{ft}{s} \tag{2}$$

Combining Eqs. 1 and 2 we get

$$p_1 = \rho\left(\frac{V_3^2 - V_1^2}{2}\right) = \left(0.00238\ \frac{slug}{ft^3}\right)\left(\frac{1\ lb \cdot s^2}{slug \cdot ft}\right)\left[\frac{\left(50\ \frac{ft}{s}\right)^2 - \left(30\ \frac{ft}{s}\right)^2}{2}\right]$$

or

$$p_1 = \underline{\underline{1.90\ \frac{lb}{ft^2}}}$$

(continued)

5.7R continued

For part (c) we determine the gage pressure at the tip of the plug, point (2), by applying the Bernoulli equation between points (1) and (2). Thus, since $V_2 = 0$,

$$\frac{p_1}{\rho} + \frac{V_1^2}{2} = \frac{p_2}{\rho}$$

or

$$p_2 = \rho\left(\frac{p_1}{\rho} + \frac{V_1^2}{2}\right)$$

$$p_2 = \left(0.00238\ \frac{\text{slug}}{\text{ft}^3}\right)\left(1\ \frac{\text{lb}\cdot\text{s}^2}{\text{slug}\cdot\text{ft}}\right)\left[\frac{1.90\ \frac{\text{lb}}{\text{ft}^2}}{\left(0.00238\ \frac{\text{slug}}{\text{ft}^3}\right)\left(1\ \frac{\text{lb}\cdot\text{s}^2}{\text{slug}\cdot\text{ft}}\right)} + \frac{\left(30\ \frac{\text{ft}}{\text{s}}\right)^2}{2}\right]$$

and

$$p_2 = \underline{\underline{2.97}}\ \frac{\text{lb}}{\text{ft}^2}$$

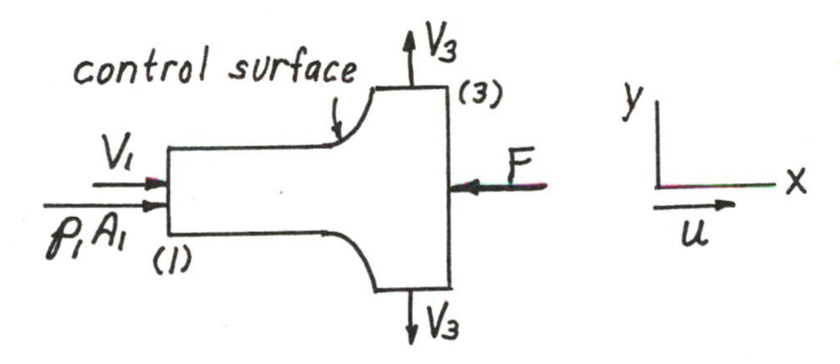

For part (d) we apply the linear momentum equation to the contents of the control volume sketched above to get

$$\int_{cs} u\rho\vec{V}\cdot\hat{n}\,dA = \Sigma F_x \quad \text{or} \quad \int_{(1)} u\rho\vec{V}\cdot dA = \Sigma F_x \quad \text{since } u_3 = 0$$

Thus,

$$-V_1\rho V_1 A_1 = p_1 A_1 - F \quad \text{or} \quad F = p_1 A_1 + V_1 \rho Q$$

Thus,

$$F = \left(1.90\frac{\text{lb}}{\text{ft}^2}\right)\frac{\pi(1\ \text{ft})^2}{4} + \left(30\ \frac{\text{ft}}{\text{s}}\right)\left(0.00238\frac{\text{slug}}{\text{ft}^3}\right)\left(1\ \frac{\text{lb}\cdot\text{s}^2}{\text{slug}\cdot\text{ft}}\right)\left(23.6\ \frac{\text{ft}^3}{\text{s}}\right)$$

$$F = \underline{\underline{3.18}}\ \text{lb}$$

5.8R

5.8R (Linear momentum) A nozzle is attached to an 80-mm inside-diameter flexible hose. The nozzle area is 500 mm^2. If the delivery pressure of water at the nozzle inlet is 700 kPa, could you hold the hose and nozzle stationary? Explain.

(ANS: yes, 707 N or 159 lb)

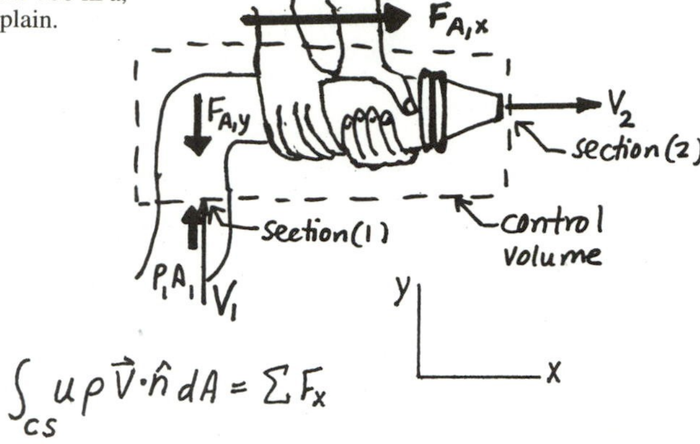

$$\int_{cs} u\rho \vec{V}\cdot\hat{n}\, dA = \sum F_x$$

The control volume shown in the sketch is used. We assume that the vertical component of the anchoring force, $F_{A,y}$, is exerted by the hose material. We further assume that the horizontal component of the anchoring force, $F_{A,x}$, must be exerted by the hands holding the hose and nozzle stationary. Application of the horizontal or x direction component of the linear momentum equation leads to

$$V_2 \rho V_2 A_2 = F_{A,x} \qquad (1)$$

Application of Bernoulli's equation between sections (1) and (2) yields

$$\frac{p_1}{\rho} + \frac{V_1^2}{2} = \frac{p_2}{\rho} + \frac{V_2^2}{2} \qquad (2)$$

From the conservation of mass equation

$$\rho V_1 A_1 = \rho V_2 A_2$$

or

$$V_1 = V_2 \frac{A_2}{A_1} \qquad (3)$$

Thus combining Eqs. 2 and 3 gives

$$V_2^2 = \frac{2(p_1 - p_2)}{\rho\left[1 - \left(\frac{A_2}{A_1}\right)^2\right]}$$

and Eq. 1 becomes

$$F_{A,x} = \frac{2(p_1 - p_2)A_2}{\left[1 - \left(\frac{A_2}{A_1}\right)^2\right]}$$

(continued)

5.8R continued

Now
$$\frac{A_2}{A_1} = \frac{(500\ mm^2)}{\left(\frac{\pi D_1^2}{4}\right)} = \frac{(500\ mm^2)}{\frac{\pi (80mm)^2}{4}} = 0.0995$$

Thus,

$$F_{A,x} = \frac{2(700\ kPa - 0\ kPa)(500\ mm^2)}{[1-(0.0995)^2]\left(1000\ \frac{mm}{m}\right)^2}\left(1000\ \frac{N}{m^2\ kPa}\right)$$

$$F_{A,x} = 707\ N$$

or in terms of lb

$$F_{A,x} = \frac{707\ N}{4.448\ \frac{N}{lb}} = \underline{\underline{159\ lb}}$$

which is managable.

5.9R

5.9R (Linear momentum) A horizontal air jet having a velocity of 50 m/s and a diameter of 20 mm strikes the inside surface of a hollow hemisphere as indicated in Fig. P5.9R. How large is the horizontal anchoring force needed to hold the hemisphere in place? The magnitude of velocity of the air remains constant.

(ANS: 1.93 N)

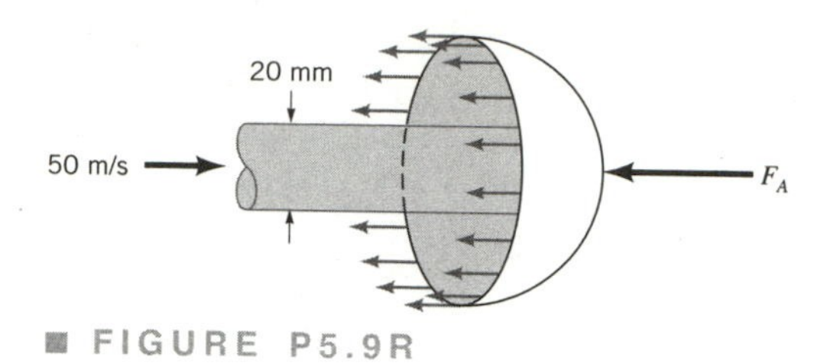

■ FIGURE P5.9R

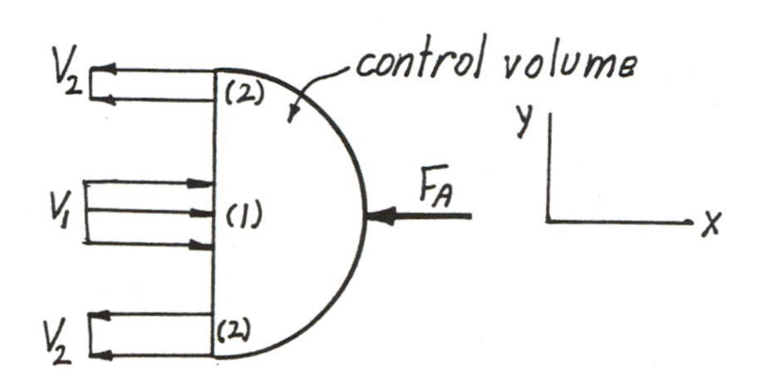

The control volume shown in the sketch is used. The x-component of the momentum equation gives

$$\int_{cs} u \rho \vec{V} \cdot \hat{n} \, dA = \sum F_x \quad \text{or}$$

$$(V_1)\rho_1(-V_1)A_1 + (-V_2)\rho_2(+V_2)A_2 = -F_A \,,$$

where for conservation of mass

$$\rho_1 A_1 V_1 = \rho_2 A_2 V_2 = \dot{m}, \text{ the mass flowrate.}$$

Thus,

$$F_A = \dot{m}(V_1 + V_2) = 2\dot{m}V \text{ since } V_1 = V_2 = 50\,\frac{m}{s}$$

Note that $V_1 = V_2$ (i.e. the speed is constant), but $\vec{V}_1 = 50\hat{i}\,\frac{m}{s} \neq \vec{V}_2 = -50\hat{i}\,\frac{m}{s}$ (i.e. the velocity changes).

With

$$\dot{m} = \rho_1 A_1 V_1 = 1.23\,\frac{kg}{m^3}\left(\frac{\pi}{4}(0.020\,m)^2\right)\left(50\,\frac{m}{s}\right) = 0.0193\,\frac{kg}{s}$$

we obtain

$$F_A = \left(0.0193\,\frac{kg}{s}\right)\left(50\,\frac{m}{s}\right)(2) = \underline{\underline{1.93\,N}}$$

5.10R

5.10R (Linear momentum) Determine the magnitude of the horizontal component of the anchoring force required to hold in place the 10-foot-wide sluice gate shown in Fig. P5.10R. Compare this result with the size of the horizontal component of the anchoring force required to hold in place the sluice gate when it is closed and the depth of water upstream is 6 ft.

(ANS: 5310 lb; 11,200 lb)

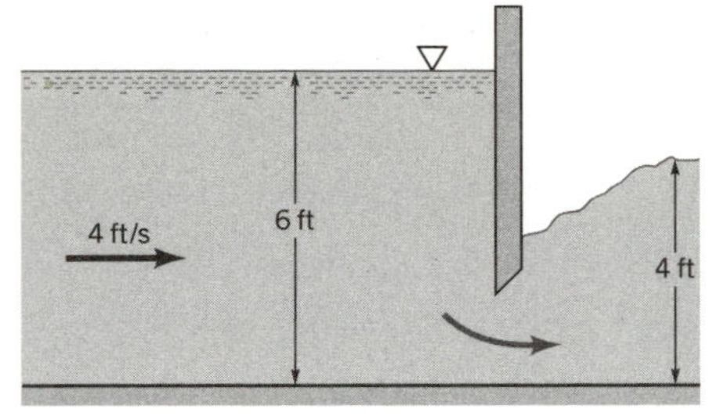

■ FIGURE P5.10R

When the gate is closed the water is stationary and the resultant water force on the gate is

$$R_{x_{closed}} = p_c A = \gamma h_c A = \gamma \frac{H^2}{2} b \quad \text{where } H = 6\,ft \text{ and } b = 10\,ft = \text{width}$$

Thus,

$$R_{x_{closed}} = \frac{1}{2}\left(62.4\,\frac{lb}{ft^3}\right)(6\,ft)^2(10\,ft) = \underline{\underline{11{,}200\,lb}}$$

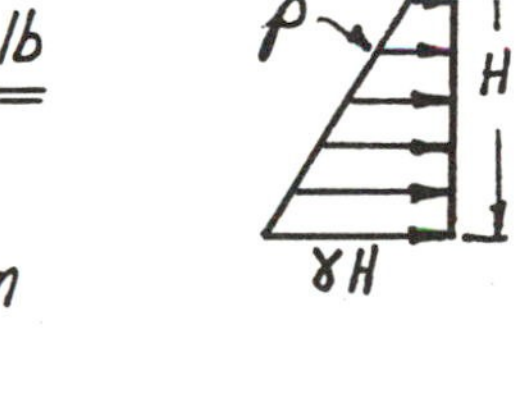

From the control volume diagram shown, the x-component of the momentum equation becomes:

$$\int_{cs} u \rho \vec{V} \cdot \hat{n}\, dA = \Sigma F_x \quad \text{or}$$

$$V_1 \rho(-V_1)A_1 + V_2 \rho(+V_2)A_2 = p_{c_1}A_1 - p_{c_2}A_2 - R_{x_{open}}$$

or

$$R_{x_{open}} = p_{c_1}A_1 - p_{c_2}A_2 + \rho V_1^2 A_1 - \rho V_2^2 A_2$$

$$= \frac{1}{2}\gamma H^2 b - \frac{1}{2}\gamma h^2 b + \rho V_1^2 H b - \rho V_2^2 h b$$

where

$$V_1 = 4\,\frac{ft}{s} \text{ and } V_2 = \frac{V_1 A_1}{A_2} = V_1 \frac{H}{h} = 4\,\frac{ft}{s}\left(\frac{6\,ft}{4\,ft}\right) = 6\,\frac{ft}{s}$$

Thus,

$$R_{x_{open}} = \frac{1}{2}\left(62.4\,\frac{lb}{ft^3}\right)(6\,ft)^2(10\,ft) - \frac{1}{2}\left(62.4\,\frac{lb}{ft^3}\right)(4\,ft)^2(10\,ft)$$
$$+ 1.94\,\frac{slugs}{ft^3}\left(4\,\frac{ft}{s}\right)^2(6\,ft)(10\,ft) - 1.94\,\frac{slugs}{ft^3}\left(6\,\frac{ft}{s}\right)^2(4\,ft)(10\,ft)$$

or

$$R_{x_{open}} = \underline{\underline{5{,}310\,lb}} < R_{x_{closed}} = 11{,}200\,lb$$

5.11R

5.11R (Linear momentum) Two jets of liquid, one with specific gravity 1.0 and the other with specific gravity 1.3, collide and form one homogeneous jet as shown in Fig. P5.11R. Determine the speed, V, and the direction, θ, of the combined jet. Gravity is negligible.

(ANS: 6.97 ft/s; 70.3 deg)

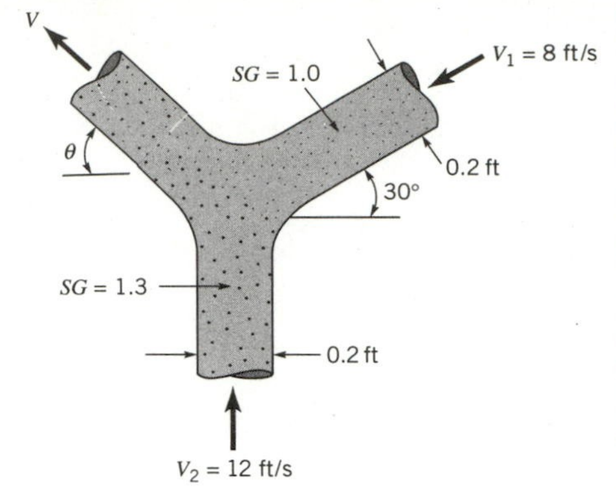

■ FIGURE P5.11R

For the control volume shown, the x-component of the momentum equation becomes $\int_{cs} u\rho\vec{V}\cdot\hat{n}\,dA = \Sigma F_x$, or

$$(-V_1\cos 30^\circ)\rho_1(-V_1)A_1 - (V\cos\theta)\rho(V)A = 0 \qquad (1)$$

since there is no force acting on the control volume. Similarly, in the y-direction

$\int_{cs} v\rho\vec{V}\cdot\hat{n}\,dA = \Sigma F_y$, or

$$(-V_1\sin 30^\circ)\rho_1(-V_1)A_1 + V_2\rho_2(-V_2)A_2 + (V\sin\theta)\rho(V)A = 0 \qquad (2)$$

since $\Sigma F_y = 0$

By combining Eqs. (1) and (2) we obtain (divide (1) by (2)):

$$\cot\theta = \frac{\rho_1 V_1^2\cos 30^\circ A_1}{\rho_2 V_2^2 A_2 - \rho_1 V_1^2 \sin 30^\circ A_1} \quad \text{where } \rho_2 = 1.3\rho_1$$

Thus,

$$\cot\theta = \frac{\rho_1\left(8\frac{ft}{s}\right)^2\cos 30^\circ\frac{\pi}{4}(0.2\,ft)^2}{(1.3\rho_1)\left(12\frac{ft}{s}\right)^2\frac{\pi}{4}(0.2\,ft)^2 - \rho_1\left(8\frac{ft}{s}\right)^2\frac{\pi}{4}(0.2\,ft)^2\sin 30^\circ} = 0.357$$

so that

$$\theta = \underline{\underline{70.3^\circ}}$$

Also, for conservation of mass: $\rho VA = \rho_1 V_1 A_1 + \rho_2 V_2 A_2 \qquad (3)$

By combining Eqs. (1) and (3)

$\rho_1 V_1^2\cos 30^\circ A_1 - V\cos\theta\left[\rho_1 V_1 A_1 + \rho_2 V_2 A_2\right] = 0$ or since $A_1 = A_2$

$$V = \frac{\rho_1 V_1^2\cos 30^\circ}{\cos\theta(\rho_1 V_1 + \rho_2 V_2)} = \frac{\rho_1\left(8\frac{ft}{s}\right)^2\cos 30^\circ}{(\cos 70.3^\circ)\left[\rho_1\left(8\frac{ft}{s}\right) + 1.3\rho_1\left(12\frac{ft}{s}\right)\right]} = \underline{\underline{6.97\,\frac{ft}{s}}}$$

5.12R

5.12R (Linear momentum) Water flows vertically upward in a circular cross-sectional pipe as shown in Fig. P5.12R. At section (1), the velocity profile over the cross-sectional area is uniform. At section (2), the velocity profile is

$$\mathbf{V} = w_c\left(\frac{R-r}{R}\right)\hat{\mathbf{k}}$$

where $\mathbf{V}$ = local velocity vector, w_c = centerline velocity in the axial direction, R = pipe radius, and r = radius from pipe axis. Develop an expression for the fluid pressure drop that occurs between sections (1) and (2).

(ANS: $p_1 - p_2 = R_z/\pi R^2 + 0.50\,\rho w_1^2 + g\rho h$, where R_z = friction force)

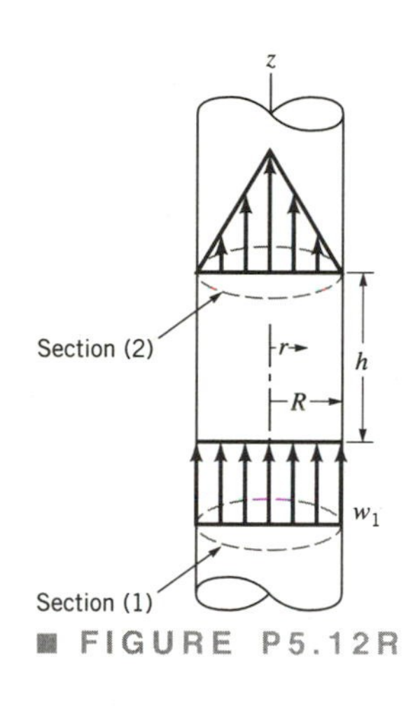

■ FIGURE P5.12R

R_z = axial force of pipe wall on the fluid

W_w = weight of water

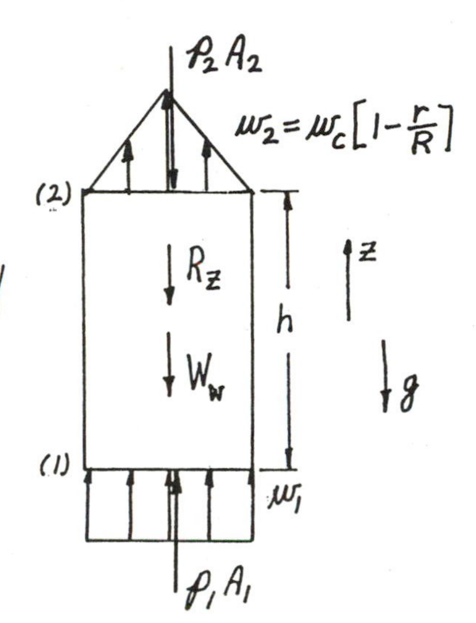

For the control volume shown in the figure the z-component of the momentum equation is

$$\int_{cs} w\rho\vec{V}\cdot\hat{n}\,dA = \Sigma F_z \quad \text{or}$$

$$-w_1\rho w_1 A_1 + \int_{r=0}^{R} w_2\rho w_2\,(2\pi r\,dr) = p_1A_1 - p_2A_2 - R_z - W_w$$

where $dA = 2\pi r\,dr$

Thus, with $A_1 = A_2 \equiv A$ this becomes

$$p_1 - p_2 = \frac{R_z}{A} + \frac{W_w}{A} - \rho w_1^2 + \frac{2\pi\rho}{A}\int_0^R \left[w_c\left(1-\frac{r}{R}\right)\right]^2 r\,dr \qquad (1)$$

But with $x \equiv \frac{r}{R}$,

$$\int_0^R w_c^2\left(1-\frac{r}{R}\right)^2 r\,dr = w_c^2\int_{r=0}^{R}\left(1-2\frac{r}{R}+\frac{r^2}{R^2}\right)r\,dr = w_c^2R^2\int_{x=0}^{x=1}(x-2x^2+x^3)\,dx$$

$$= w_c^2R^2\left[\frac{x^2}{2} - \frac{2}{3}x^3 + \frac{1}{4}x^4\right]_0^1 = \frac{1}{12}R^2(w_c^2)$$

(continued)

Thus, since $A=\pi R^2$, Eq. (1) becomes

$$p_1-p_2=\frac{R_z}{\pi R^2}+\frac{W_w}{\pi R^2}-\rho w_1^2+\frac{2\pi\rho}{\pi R^2}\left(\frac{1}{12}R^2\right)w_c^2 \qquad (2)$$

We can determine w_c in terms of w_1 by using the continuity equation: $\int_{cs}\rho\vec{V}\cdot\hat{n}\,dA=0$, or since $\rho\equiv$ constant:

$$A_1w_1=\int w_2\,dA=\int_{r=0}^{R}w_c\left[1-\frac{r}{R}\right](2\pi r\,dr)=2\pi w_c R^2\int_{x=0}^{1}(x-x^2)\,dx$$

or

$$\pi R^2 w_1=2\pi w_c R^2\left(\frac{1}{6}\right)$$

Thus,

$w_c=3w_1$ and Eq. (2) becomes

$$p_1-p_2=\frac{R_z}{\pi R^2}+\frac{W_w}{\pi R^2}-\rho w_1^2+\rho\left(\frac{1}{6}\right)(3w_1)^2$$

or

$$p_1-p_2=\frac{R_z}{\pi R^2}+\frac{W_w}{\pi R^2}+\frac{1}{2}\rho w_1^2$$

or with $W_w=\gamma A h=\rho g\pi R^2 h$

$$p_1-p_2=\frac{R_z}{\pi R^2}+\rho g h+\frac{1}{2}\rho w_1^2$$

5.13R

5.13R (Moment-of-momentum) A lawn sprinkler is constructed from pipe with $\frac{1}{4}$-in. inside diameter as indicated in Fig. P5.13R. Each arm is 6 in. in length. Water flows through the sprinkler at the rate of 1.5 lb/s. A force of 3 lb positioned halfway along one arm holds the sprinkler stationary. Compute the angle, θ, which the exiting water stream makes with the tangential direction. The flow leaves the nozzles in the horizontal plane.

(ANS: 23.9 deg)

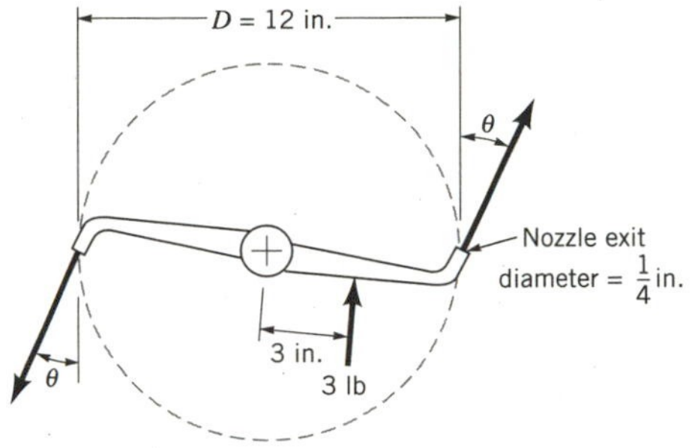

■ FIGURE P5.13R

The stationary, non-deforming control volume shown in the sketch is used. Application of the axial component of the moment-of-momentum equation (Eq. 5.50) leads to

$$T_{shaft} = \dot{m}\, r_2\, V_{\theta,2} = \dot{m}\, r_2\, V_2 \cos\theta \qquad (1)$$

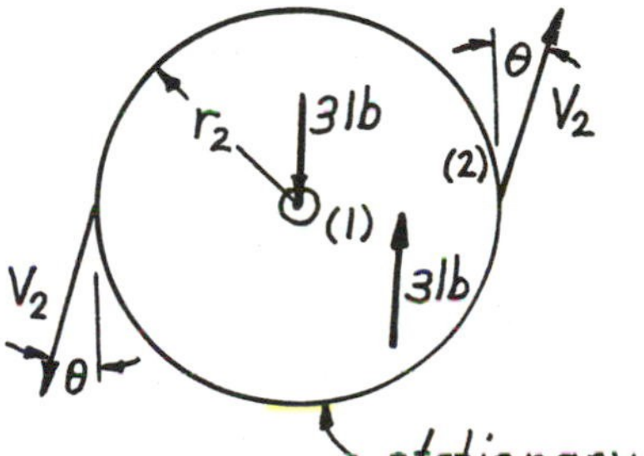

Since

$$V_2 = \frac{\dot{m}}{2\rho A_{nozzle\ opening}}$$

where

$$A_{nozzle} = \frac{\pi D^2_{nozzle\ opening}}{4}$$

Eq. 1 leads to

$$\cos\theta = \frac{T_{shaft}\, 2\rho\, \frac{\pi D^2_{nozzle\ opening}}{4}}{\dot{m}^2 r_2}$$

where $\dot{m} = \rho V_2 A_2 = \frac{\gamma}{g} Q = \frac{1.5\ lb/s}{32.2\ ft/s^2} = 0.0466\ \frac{slug}{s}$

$$\theta = \cos^{-1} \frac{(3\ lb)(\frac{3}{12} ft)(2)(1.94\ \frac{slugs}{ft^3})\frac{\pi}{4}(\frac{1}{48} ft)^2}{(0.0466\ \frac{slugs}{s})^2 (\frac{6}{12} ft)}$$

Thus,

$$\theta = \underline{\underline{23.9^\circ}}$$

5.14R

5.14R (Moment-of-momentum) A water turbine with radial flow has the dimensions shown in Fig. P5.14R. The absolute entering velocity is 15 m/s, and it makes an angle of 30° with the tangent to the rotor. The absolute exit velocity is directed radially inward. The angular speed of the rotor is 30 rpm. Find the power delivered to the shaft of the turbine.

(ANS: −7.68 MW)

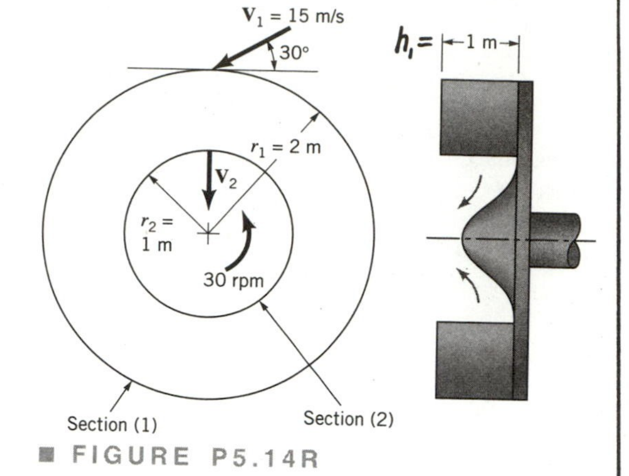

■ FIGURE P5.14R

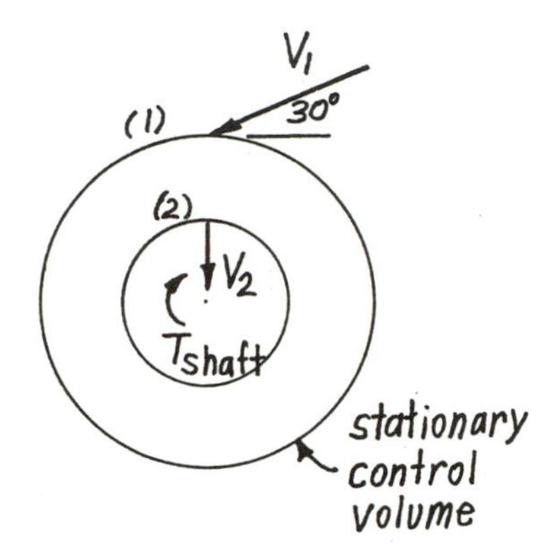

The stationary and non-deforming control volume shown in the sketch above is used. We use Eq. 5.53 to determine the shaft power involved. Thus

$$\dot{W}_{shaft} = -\dot{m}_1 U_1 V_{\theta,1} + \dot{m}_2 U_2 V_{\theta,2}\text{, where } V_{\theta,2}=0 \qquad (1)$$

The mass flowrate may be obtained from (2)

$$\dot{m}_1 = \rho V_{R,1} A_1 = \rho V_{R,1} 2\pi r_1 h_1$$

where

$$V_{R,1} = \text{radial component of velocity at section(1)}$$

The blade velocity at section (1) is

$$U_1 = r_1 \omega = (2\ m)\left(30\ \frac{rev}{min}\right)\left(2\pi\ \frac{rad}{rev}\right)\frac{1}{\left(60\ \frac{s}{min}\right)} = 6.28\ \frac{m}{s}$$

The values of $V_{\theta,1}$ and $V_{R,1}$ may be obtained with the help of a velocity triangle for the flow at section (1) as sketched below.

(continued)

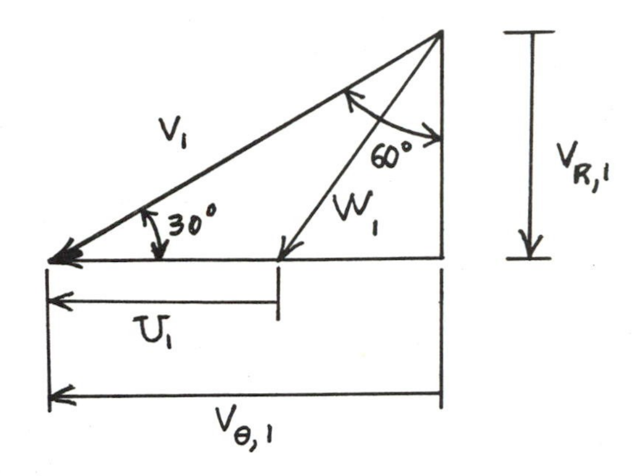

With the velocity triangle we conclude that

$$V_{R,1} = V_1 \sin 30° = V_1 \cos 60° = \left(15 \frac{m}{s}\right)(\sin 30°) = 7.5 \frac{m}{s}$$

Then from Eq. 2

$$\dot{m}_1 = \left(999 \frac{kg}{m^3}\right)\left(7.5 \frac{m}{s}\right) 2\pi (2\,m)(1\,m) = 94{,}100 \frac{kg}{s}$$

Also, with the triangle we see that

$$V_{\theta,1} = V_1 \cos 30° = V_1 \sin 60° = \left(15 \frac{m}{s}\right) \cos 30° = 13.0 \frac{m}{s}$$

Then, with Eq. 1 we obtain

$$\dot{W}_{shaft} = -\left(94{,}100 \frac{kg}{s}\right)\left(6.28 \frac{m}{s}\right)\left(13.0 \frac{m}{s}\right)\left(\frac{1\,N}{kg \cdot \frac{m}{s^2}}\right)\left(\frac{MW}{10^6 \frac{N \cdot m}{s}}\right)$$

$$\dot{W}_{shaft} = \underline{\underline{-7.68\ MW}}$$

5.15R

5.15R (Moment-of-momentum) The single stage, axial-flow turbomachine shown in Fig. P5.15R involves water flow at a volumetric flowrate of 11 m^3/s. The rotor revolves at 600 rpm. The inner and outer radii of the annular flow path through the stage are 0.46 and 0.61 m, and $\beta_2 = 30°$. The flow entering the rotor row and leaving the stator row is axial viewed from the stationary casing. Is this device a turbine or a pump? Estimate the amount of power transferred to or from the fluid.

(ANS: pump; 7760 kW)

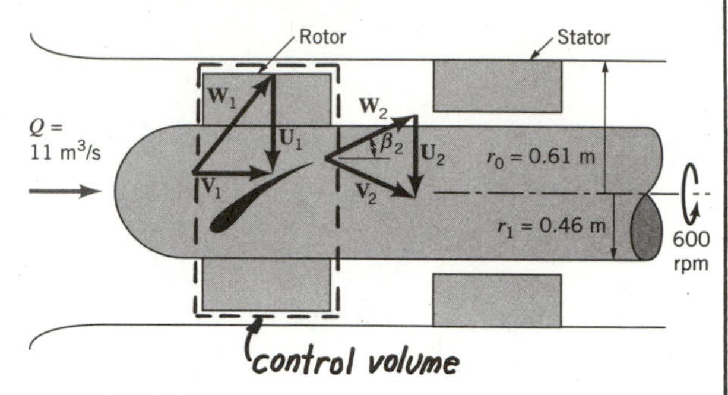

■ FIGURE P5.15R

This device is a pump because the lift force acting on each rotor blade is opposite in direction to the blade motion. The direction of the blade lift force is ascertained by noting how the blade turns the flow past the blade. The power transferred from the rotor blades to the water may be evaluated with the moment-of-momentum power equation (equation (5.53) with $V_{\theta,1}=0$):

$$\dot{W}_{\substack{shaft\\net\ in}} = \dot{m}\, U_2 V_{\theta,2} = \rho Q U_2 V_{\theta,2} \qquad (1)$$

We obtain U_2 with

$$U_2 = r_2\omega = \left(\frac{r_i + r_o}{2}\right)\omega = \left(\frac{0.46\,m + 0.61\,m}{2}\right)\frac{(600\,rpm)}{\left(60\,\frac{s}{min}\right)}\left(2\pi\,\frac{rad}{rev}\right) = 33.6\,\frac{m}{s}$$

From the rotor exit flow velocity triangle we obtain $V_{\theta,2}$ with

$$V_{\theta,2} = U_2 - V_{axial,2}\tan\beta_2 \quad \text{(see figure below)}$$

where

$$V_{axial,2} = \frac{Q}{A_2} = \frac{Q}{\pi(r_o^2 - r_i^2)} = \frac{\left(11\,\frac{m^3}{s}\right)}{\pi\left[(0.61m)^2 - (0.46m)^2\right]} = 21.8\,\frac{m}{s}$$

Thus

$$V_{\theta,2} = 33.6\,\frac{m}{s} - \left(21.8\,\frac{m}{s}\right)\tan 30° = 21.0\,\frac{m}{s}$$

and with Eq. 1 we get

$$\dot{W}_{\substack{shaft\\net\ in}} = \left(999\,\frac{kg}{m^3}\right)\left(11\,\frac{m^3}{s}\right)\left(33.6\,\frac{m}{s}\right)\left(21.0\,\frac{m}{s}\right)\left(\frac{1\,N}{kg\cdot\frac{m}{s^2}}\right) = 7.76\times10^6\,\frac{N\cdot m}{s}$$

or

$$\dot{W}_{\substack{shaft\\net\ in}} = \underline{\underline{7760\,kW}}$$

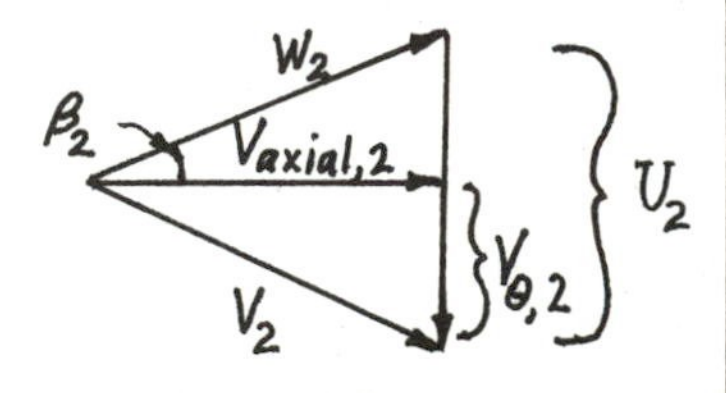

Note: Since $\dot{W}_{\substack{shaft\\net\ in}} > 0$ this device is a pump.

5.16R

5.16R (Moment-of-momentum) A small water turbine is designed as shown in Fig. P5.16R. If the flowrate through the turbine is 0.0030 slugs/s, and the rotor speed is 300 rpm, estimate the shaft torque and shaft power involved. Each nozzle exit cross-sectional area is 3.5×10^{-5} ft^2.
(ANS: -0.0107 ft·lb; -0.336 ft·lb/s)

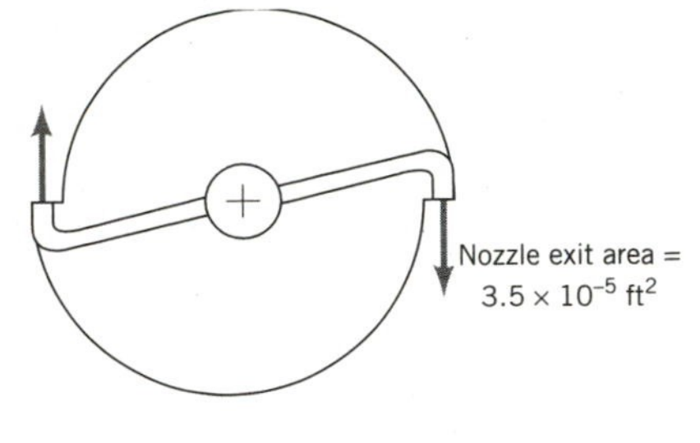

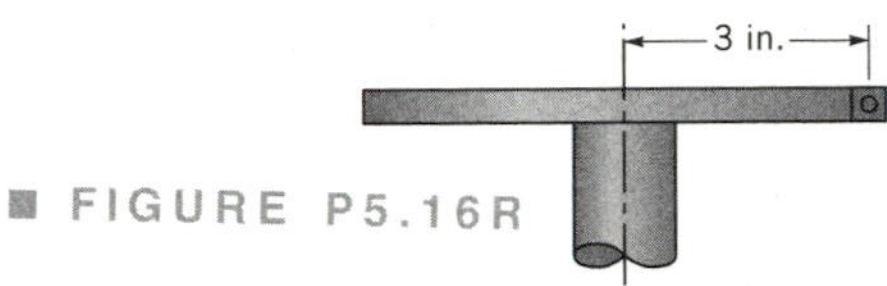

■ FIGURE P5.16R

For shaft torque we can use the axial component of the moment-of-momentum equation (Eq. 5.50). Thus, with $V_{\theta,1} = 0$:

$$T_{shaft} = -\dot{m}_2 r_2 V_{\theta,2}$$

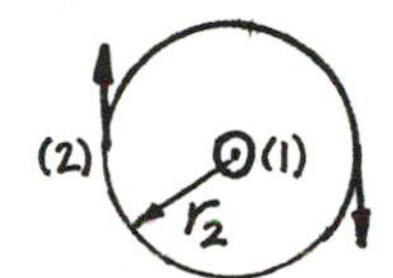

Consideration of the absolute and relative velocities of the flow out of each nozzle (see sketch below) leads to

$$V_{\theta,2} = V_2 = W_2 - U_2$$

$V_2 = V_{\theta,2}$

W_2

U_2

where

$$W_2 = \frac{\dot{m}}{2\rho A_{\substack{nozzle\\exit}}}$$

and

$$U_2 = r_2 \omega$$

Thus,

$$T_{shaft} = -\dot{m}_2 r_2 \left(\frac{\dot{m}}{2\rho A_{\substack{nozzle\\exit}}} - r_2 \omega \right)$$

$$T_{shaft} = \frac{-(0.003 \frac{slugs}{ft^3})(3 \text{ in.})}{(12 \frac{in.}{ft})} \left[\frac{0.003 \frac{slugs}{ft^3}}{2(1.94 \frac{slugs}{ft^3})(3.5 \times 10^{-5} ft^2)} - \frac{(3 \text{ in.})(300 \frac{rev}{min})(2\pi \frac{rad}{rev})}{(12 \frac{in.}{ft})(60 \frac{s}{min})} \right] \left(\frac{1 \text{ lb}}{slug \cdot \frac{ft}{s^2}} \right)$$

(continued)

5.16 continued

or $T_{shaft} = -0.0107$ ft·lb (minus sign mean torque opposes rotation)

Now,

$$\dot{W}_{shaft} = T_{shaft}\,\omega$$

or

$$\dot{W}_{shaft} = \frac{(-0.0107 \text{ ft}\cdot\text{lb})\left(300 \frac{\text{rev}}{\text{min}}\right)\left(2\pi \frac{\text{rad}}{\text{rev}}\right)}{\left(60 \frac{\text{s}}{\text{min}}\right)} = -0.336 \frac{\text{ft}\cdot\text{lb}}{\text{s}}$$

(minus sign means work is out of the control volume)

5.17R

5.17R (Energy equation) Water flows steadily from one location to another in the inclined pipe shown in Fig. P5.17R. At one section, the static pressure is 12 psi. At the other section, the static pressure is 5 psi. Which way is the water flowing? Explain.

(ANS: from A to B)

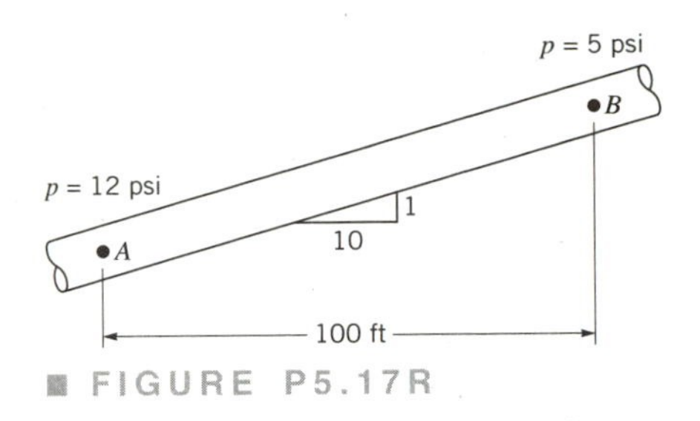

■ FIGURE P5.17R

To determine the direction of water flow we apply the energy equation (Eq. 5.82) for flow from sections (A) to (B) and flow from sections (B) to (A). The loss obtained with Eq. 5.82 is positive for the correct flow direction, but negative for the incorrect flow direction.

For flow from sections (A) to (B), Eq. 5.82 leads to

$$loss = \frac{p_A - p_B}{\rho} + \frac{V_A^2 - V_B^2}{2} + g(z_A - z_B) + w_{shaft\ net\ in}$$

(with $\frac{V_A^2 - V_B^2}{2} = 0$ and $w_{shaft\ net\ in} = 0$)

or

$$loss = \frac{(12\,psi - 5\,psi)\left(144\,\frac{in.^2}{ft^2}\right)}{\left(1.94\,\frac{slugs}{ft^3}\right)} + \left(32.2\,\frac{ft}{s^2}\right)(-10\,ft)\left(\frac{1\,lb}{slug\cdot\frac{ft}{s^2}}\right)$$

and

$$loss = 198\,\frac{ft\cdot lb}{slug}$$

For flow from sections (B) to (A), Eq. 5.82 leads to

$$loss = \frac{p_B - p_A}{\rho} + g(z_B - z_A)$$

or

$$loss = \frac{(5\,psi - 12\,psi)\left(144\,\frac{in.^2}{ft^2}\right)}{\left(1.94\,\frac{slugs}{ft^3}\right)} + \left(32.2\,\frac{ft}{s^2}\right)(10\,ft)\left(\frac{1\,lb}{slug\cdot\frac{ft}{s^2}}\right)$$

and

$$loss = -198\,\frac{ft\cdot lb}{slug}$$

The water flow is from <u>section (A) to section (B)</u> (i.e., uphill)

5.18R

5.18R (Energy equation) The pump shown in Fig. P5.18R adds 20 kW of power to the flowing water. The only loss is that which occurs across the filter at the inlet of the pump. Determine the head loss for this filter.

(ANS: 7.69 m)

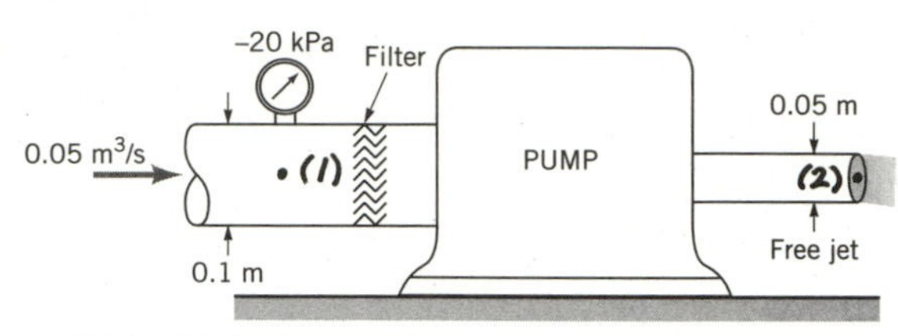

■ FIGURE P5.18R

The energy equation for this flow can be written as

$$\frac{p_1}{\gamma} + \frac{V_1^2}{2g} + z_1 + h_p = \frac{p_2}{\gamma} + \frac{V_2^2}{2g} + z_2 + h_L \qquad (1)$$

where

$z_1 = z_2$, $p_2 = 0$, $p_1 = -20\ kPa$, and

$$V_1 = \frac{Q}{A_1} = \frac{0.05\ \frac{m^3}{s}}{\frac{\pi}{4}(0.1\ m)^2} = 6.37\ \frac{m}{s}$$

$$V_2 = \frac{Q}{A_2} = \frac{0.05\ \frac{m^3}{s}}{\frac{\pi}{4}(0.05\ m)^2} = 25.5\ \frac{m}{s}$$

Also,

$$h_p = \frac{\dot{W}_s}{\gamma Q} = \frac{20\times10^3\ \frac{N\cdot m}{s}}{(9.8\times10^3\ \frac{N}{m^3})(0.05\ \frac{m^3}{s})} = 40.8\ m$$

Thus, Eq. (1) becomes

$$\frac{(-20\times10^3\ \frac{N}{m^2})}{(9.8\times10^3\ \frac{N}{m^3})} + \frac{(6.37\ \frac{m}{s})^2}{2(9.81\ \frac{m}{s^2})} + 40.8\ m = \frac{(25.5\ \frac{m}{s})^2}{2(9.81\ \frac{m}{s^2})} + h_L$$

or

$$h_L = \underline{\underline{7.69\ m}}$$

5.19R

5.19R (Linear momentum/energy) Eleven equally spaced turning vanes are used in the horizontal plane 90° bend as indicated in Fig. P5.19R. The depth of the rectangular cross-sectional bend remains constant at 3 in. The velocity distributions upstream and downstream of the vanes may be considered uniform. The loss in available energy across the vanes is $0.2V_1^2/2$. The required velocity and pressure downstream of the vanes, section (2), are 180 ft/s and 15 psia. What is the average magnitude of the force exerted by the air flow on each vane? Assume the force of the air on the duct walls is equivalent to the force of the air on one vane.

(ANS: 4.61 lb)

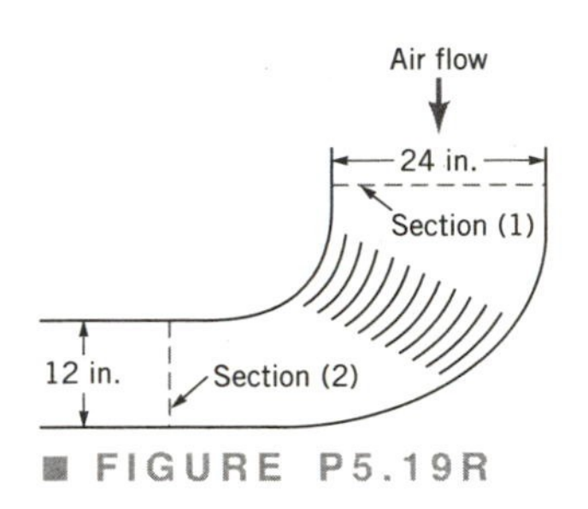

■ FIGURE P5.19R

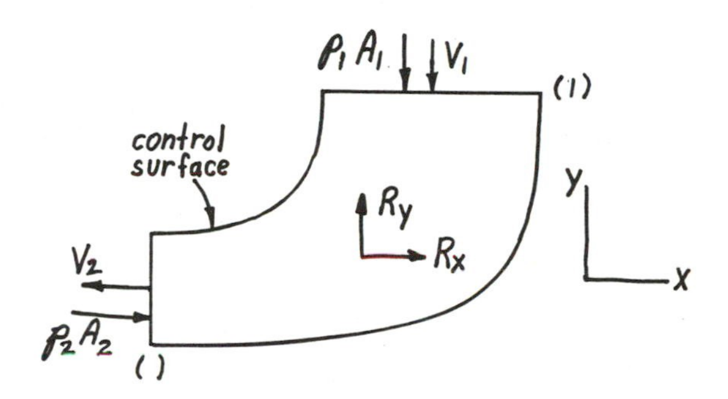

To estimate the average magnitude of the force exerted by the air flow on each vane, we determine the magnitude of the resultant force exerted by the air on the vanes and the duct walls and divide that result by 12. We assume that the duct walls act as one additional vane. The linear momentum equation (Eq. 5.22) is used to determine the x and y components of the resultant force exerted by the vanes and duct walls on the air between sections (1) and (2). Thus, $\sum \vec{F} = \int_{cs} \vec{V} \rho \vec{V} \cdot \hat{n} \, dA$, or

$$R_x = -p_2 A_2 - V_2 \rho A_2 V_2 \quad (1)$$

and

$$R_y = p_1 A_1 + V_1 \rho A_1 V_1 \quad (2)$$

From the conservation of mass principle (Eq. 5.13) we have

$$V_1 = V_2 \frac{A_2}{A_1} = \left(180 \frac{ft}{s}\right) \frac{(12 \text{ in.})(3 \text{ in.})}{(24 \text{ in})(3 \text{ in.})} = 90 \frac{ft}{s}$$

(continued)

With the energy equation (Eq. 5.83) we obtain

$$p_1 = p_2 + \frac{\rho}{2}\left(V_2^2 - V_1^2\right) + \rho(\text{loss}) = p_2 + \frac{\rho}{2}\left(V_2^2 - V_1^2 + 0.2V_1^2\right)$$

or

$$p_1 = p_2 + \frac{\rho}{2}\left(V_2^2 - 0.8V_1^2\right) = 15\,\text{psia} + \frac{\left(2.38\times10^{-3}\,\frac{\text{slug}}{\text{ft}^3}\right)}{2\left(144\,\frac{\text{in.}^2}{\text{ft}^2}\right)}\left[\left(180\,\frac{\text{ft}}{\text{s}}\right)^2 - 0.8\left(90\,\frac{\text{ft}}{\text{s}}\right)^2\right]\left(\frac{1\,\text{lb}}{\text{slug}\cdot\frac{\text{ft}}{\text{s}^2}}\right)$$

and

$$p_1 = 15.21\,\text{psia}$$

As suggested in Section 5.2.2, we use gage pressures at sections (1) and (2). Thus, from Eq 1 we have

$$R_x = -(15\,\text{psia} - 14.7\,\text{psia})(12\,\text{in.})(3\,\text{in.}) - \left(180\,\frac{\text{ft}}{\text{s}}\right)\left(2.38\times10^{-3}\,\frac{\text{slug}}{\text{ft}^3}\right)\frac{(12\,\text{in.})(3\,\text{in.})\left(180\,\frac{\text{ft}}{\text{s}}\right)}{144\,\frac{\text{in.}^2}{\text{ft}^2}}\left(\frac{1\,\text{lb}}{\text{slug}\cdot\frac{\text{ft}}{\text{s}^2}}\right)$$

or

$$R_x = -30.1\,\text{lb}$$

From Eq. 2 we obtain

$$R_y = (15.21\,\text{psia} - 14.7\,\text{psia})(24\,\text{in.})(3\,\text{in.}) + \left(90\,\frac{\text{ft}}{\text{s}}\right)\left(2.38\times10^{-3}\,\frac{\text{slug}}{\text{ft}^3}\right)\frac{(24\,\text{in.})(3\,\text{in.})\left(90\,\frac{\text{ft}}{\text{s}}\right)}{\left(144\,\frac{\text{in.}^2}{\text{ft}^2}\right)}\left(\frac{1\,\text{lb}}{\text{slug}\cdot\frac{\text{ft}}{\text{s}^2}}\right)$$

or

$$R_y = 46.4\,\text{lb}$$

Then

$$R = \sqrt{R_x^2 + R_y^2} = \sqrt{(-30.1\,\text{lb})^2 + (46.4\,\text{lb})^2} = 55.3\,\text{lb}$$

and

$$R_{\substack{\text{vane}\\\text{average}}} = \frac{R_{\text{average}}}{12} = \frac{55.3\,\text{lb}}{12} = \underline{\underline{4.61\,\text{lb}}}$$

5.20R

5.20R (Energy equation) A hydroelectric power plant operates under the conditions illustrated in Fig. P5.20R. The head loss associated with flow from the water level upstream of the dam, section (1), to the turbine discharge at atmospheric pressure, section (2), is 20 m. How much power is transferred from the water to the turbine blades?

(ANS: 23.5 MW)

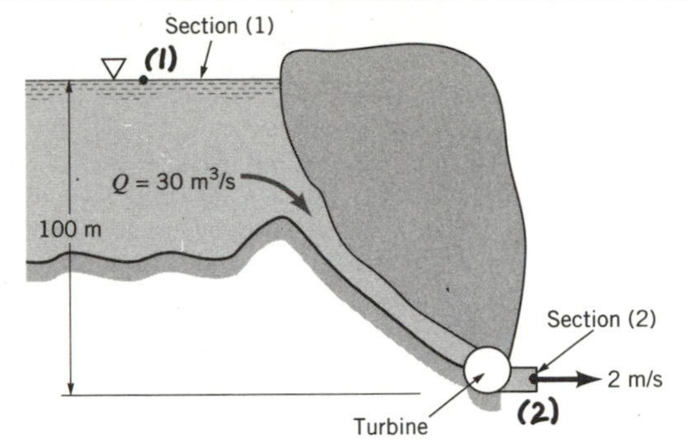

■ FIGURE P5.20R

For flow from section (1) to section (2), Eq. 5.82 leads to

$$w_{\substack{shaft\\net\ out}} = g(z_1 - z_2) - \frac{V_2^2}{2} - loss \qquad (1)$$

since

$$p_2 = p_1 = p_{atm}, \quad V_1 = 0,$$

and

$$w_{\substack{shaft\\net\ out}} = -w_{\substack{shaft\\net\ in}}$$

For power, we multiply Eq. 1 by the mass flowrate, $\dot{m}$, to get

$$\dot{W}_{\substack{shaft\\net\ out}} = \dot{m}g(z_1 - z_2) - \dot{m}\frac{V_2^2}{2} - \dot{m}\ loss$$

But

$$\dot{m} = \rho Q$$

thus

$$\dot{W}_{\substack{shaft\\net\ out}} = \rho Q g(z_1 - z_2) - \rho Q\frac{V_2^2}{2} - \rho Q\ loss$$

or

$$\dot{W}_{\substack{shaft\\net\ out}} = \left(999\frac{kg}{m^3}\right)\left(30\frac{m^3}{s}\right)\left(9.81\frac{m}{s^2}\right)(100\ m)\left(1\frac{N}{kg\cdot\frac{m}{s^2}}\right) - \left(999\frac{kg}{m^3}\right)\left(30\frac{m^3}{s}\right)\frac{\left(2\frac{m}{s}\right)^2}{2}\left(1\frac{N}{kg\cdot\frac{m}{s^2}}\right) - \left(999\frac{kg}{m^3}\right)\left(30\frac{m^3}{s}\right)(20m)\left(9.81\frac{m}{s^2}\right)\left(1\frac{N}{kg\cdot\frac{m}{s^2}}\right)$$

and

$$\dot{W}_{\substack{shaft\\net\ out}} = 23.5 \times 10^6\ \frac{N\cdot m}{s} = \underline{\underline{23.5\ MW}}$$

5.21R

5.21R (Energy equation) A pump transfers water from one large reservoir to another as shown in Fig. P5.21R*a*. The difference in elevation between the two reservoirs is 100 ft. The friction head loss in the piping is given by $K_L\overline{V}^2/2g$, where $\overline{V}$ is the average fluid velocity in the pipe and K_L is the loss coefficient, which is considered constant. The relation between the total head rise, H, across the pump and the flowrate, Q, through the pump is given in Fig. 5.21R*b*. If $K_L = 40$, and the pipe diameter is 4 in., what is the flowrate through the pump?

(ANS: 0.653 ft^3/s)

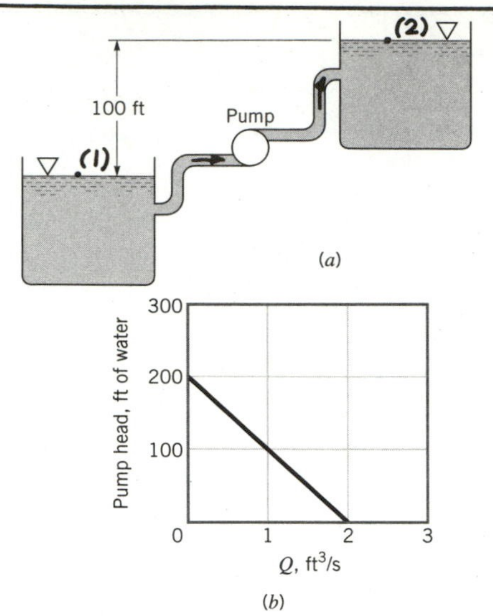

■ FIGURE P5.21R

For the flow from section (1) to section (2), Eq. 5.84 leads to

$$h_p = z_2 - z_1 + h_L \quad \text{since } V_1 = V_2 = 0 \qquad (1)$$

From Fig. P5.117 b we conclude that

$$h_p = 200 - 100\,Q \qquad (2)$$

From the problem statement

$$h_L = K_L \frac{\overline{V}^2}{2g}$$

or since

$$\overline{V} = \frac{Q}{A} = \frac{Q}{\frac{\pi D^2}{4}}$$

we have

$$h_L = \frac{K_L Q^2}{2g\left(\frac{\pi D^2}{4}\right)^2} = \frac{(40)\left(Q\,\frac{ft^3}{s}\right)^2}{(2)\left(32.2\,\frac{ft}{s^2}\right)\left[\frac{\pi\,(4\text{ in.})^2}{\left(12\,\frac{\text{in.}}{ft}\right)^2(4)}\right]^2} = 81.6\,Q^2 \text{ ft} \qquad (3)$$

Combining Eqs. 1, 2 and 3 we obtain

$$81.6\,Q^2 + 100\,Q - 100 = 0 \qquad (4)$$

The root of Eq. 4 that makes physical sense (i.e. $Q>0$) is

$$Q = 0.653\,\frac{ft^3}{s}$$

5.22R

5.22R (Energy equation) The pump shown in Fig. P5.22R adds 1.6 horsepower to the water when the flowrate is 0.6 ft^3/s. Determine the head loss bewteen the free surface in the large, open tank and the top of the fountain (where the velocity is zero).

(ANS: 7.50 ft)

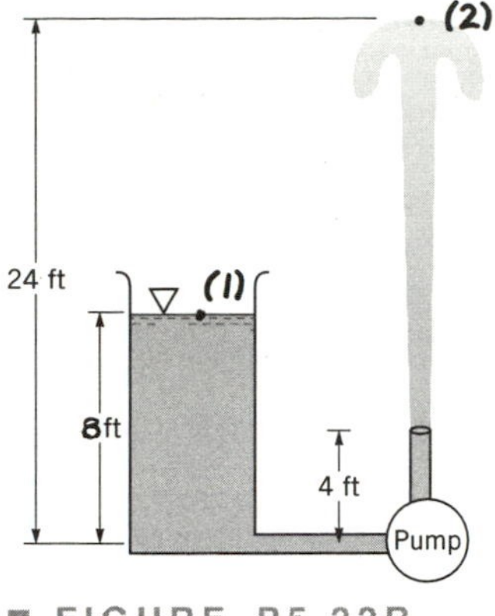

■ FIGURE P5.22R

The energy equation for this flow can be written as

$$\frac{p_1}{\gamma}+\frac{V_1^2}{2g}+z_1+h_s=\frac{p_2}{\gamma}+\frac{V_2^2}{2g}+z_2+h_L \qquad (1)$$

where

$$p_1=p_2=V_1=V_2=0$$

Also,

$$h_s=\frac{\dot{W}_s}{\gamma Q}=\frac{1.6\,hp\left(550\,\frac{ft\cdot lb}{s}/hp\right)}{\left(62.4\,\frac{lb}{ft^3}\right)\left(0.6\,\frac{ft^3}{s}\right)}=23.5\,ft$$

Thus, Eq. (1) becomes

$$h_L=z_1+h_s-z_2=8\,ft+23.5\,ft-24\,ft=\underline{\underline{7.50\,ft}}$$

Some of this head loss may occur in the pipe and some in the water jet as it interacts with the surrounding air.

6

Differential Analysis of Fluid Flow

Flow past an inclined plate: The streamlines of a viscous fluid flowing slowly past a two-dimensional object placed between two closely spaced plates (a Hele-Shaw cell) approximate inviscid, irrotational (potential) flow. (Dye in water between glass plates spaced 1 mm apart.) (Photography courtesy of D. H. Peregrine.)

6.1R

6.1R (Acceleration) The velocity in a certain flow field is given by the equation

$$\mathbf{V} = 3yz^2\hat{\mathbf{i}} + xz\hat{\mathbf{j}} + y\hat{\mathbf{k}}$$

Determine the expressions for the three rectangular components of acceleration.

(ANS: $3xz^3 + 6y^2z$; $3yz^3 + xy$; xz)

From expression for velocity, $u = 3yz^2$, $v = xz$, and $w = y$.

Since

$$a_x = \frac{\partial u}{\partial t} + u\frac{\partial u}{\partial x} + v\frac{\partial u}{\partial y} + w\frac{\partial u}{\partial z}$$

then

$$a_x = 0 + (3yz^2)(0) + (xz)(3z^2) + (y)(6yz)$$

$$= \underline{\underline{3xz^3 + 6y^2z}}$$

Similarly,

$$a_y = \frac{\partial v}{\partial t} + u\frac{\partial v}{\partial x} + v\frac{\partial v}{\partial y} + w\frac{\partial v}{\partial z}$$

and

$$a_y = 0 + (3yz^2)(z) + (xz)(0) + (y)(x)$$

$$= \underline{\underline{3yz^3 + xy}}$$

Also,

$$a_z = \frac{\partial w}{\partial t} + u\frac{\partial w}{\partial x} + v\frac{\partial w}{\partial y} + w\frac{\partial w}{\partial z}$$

so that

$$a_z = 0 + (3yz^2)(0) + (xz)(1) + (y)(0)$$

$$= \underline{\underline{xz}}$$

6.2R

6.2R (Vorticity) Determine an expression for the vorticity of the flow field described by

$$\mathbf{V} = x^2y\hat{\mathbf{i}} - xy^2\hat{\mathbf{j}}$$

Is the flow irrotational?

(ANS: $-(x^2 + y^2)\hat{\mathbf{k}}$; no)

The vorticity is twice the rotation vector:

$$\vec{\zeta} = 2\vec{\omega} = \nabla \times \vec{V} \qquad \text{(Eq. 6.17)}$$

From expression for velocity, $u = x^2y$, $v = -xy^2$, and $w = 0$, and with

$$\omega_x = \frac{1}{2}\left(\frac{\partial w}{\partial y} - \frac{\partial v}{\partial z}\right) \qquad \text{(Eq. 6.13)}$$

$$\omega_y = \frac{1}{2}\left(\frac{\partial u}{\partial z} - \frac{\partial w}{\partial x}\right) \qquad \text{(Eq. 6.14)}$$

$$\omega_z = \frac{1}{2}\left(\frac{\partial v}{\partial x} - \frac{\partial u}{\partial y}\right) \qquad \text{(Eq. 6.12)}$$

it follows that

$$\omega_x = 0, \qquad \omega_y = 0, \quad \text{and} \quad \omega_z = \frac{1}{2}\left(-y^2 - x^2\right)$$

Thus,

$$\vec{\zeta} = 2\left(\omega_x\hat{i} + \omega_y\hat{j} + \omega_z\hat{k}\right)$$

$$= 2\left[(0)\hat{i} + (0)\hat{j} + \frac{1}{2}\left(-y^2 - x^2\right)\hat{k}\right]$$

$$= \underline{\underline{-\left(x^2 + y^2\right)\hat{k}}}$$

Since $\vec{\zeta}$ is not zero everywhere, the flow is not irrotational. <u>No.</u>

6.3R

6.3R (Conservation of mass) For a certain incompressible, two-dimensional flow field the velocity component in the y direction is given by the equation

$$v = x^2 + 2xy$$

Determine the velocity component in the x direction so that the continuity equation is satisfied.

(ANS: $-x^2 + f(y)$)

To satisfy the continuity equation,

$$\frac{\partial u}{\partial x} + \frac{\partial v}{\partial y} = 0 \qquad (1)$$

Since

$$\frac{\partial v}{\partial y} = 2x$$

Then from Eq. (1)

$$\frac{\partial u}{\partial x} = -2x \qquad (2)$$

Equation (2) can be integrated with respect to x to obtain

$$\int du = -\int 2x\,dx + f(y)$$

or

$$u = \underline{\underline{-x^2 + f(y)}}$$

where $f(y)$ is an undetermined function of y.

6.4R

6.4R (Conservation of mass) For a certain incompressible flow field it is suggested that the velocity components are given by the equations

$$u = x^2 y \qquad v = 4y^3 z \qquad w = 2z$$

Is this a physically possible flow field? Explain.
(ANS: No)

Any physically possible incompressible flow field must satisfy conservation of mass as expressed by the relationship

$$\frac{\partial u}{\partial x} + \frac{\partial v}{\partial y} + \frac{\partial w}{\partial z} = 0 \qquad (1)$$

For the velocity distribution given,

$$\frac{\partial u}{\partial x} = 2xy, \quad \frac{\partial v}{\partial y} = 12y^2 z, \text{ and } \frac{\partial w}{\partial z} = 2$$

Substitution into Eq. (1) shows that

$$2xy + 12y^2 z + 2 \neq 0 \text{ for all } x, y, z.$$

Thus, this is not a physically possible flow field. No.

6.5R (Stream function) The velocity potential for a certain flow field is

$$\phi = 4xy$$

Determine the corresponding stream function.
(ANS: $2(y^2 - x^2) + C$)

For the given velocity potential,

$$u = \frac{\partial \phi}{\partial x} = 4y \quad \text{and} \quad v = \frac{\partial \phi}{\partial y} = 4x$$

From the definition of the stream function,

$$u = \frac{\partial \psi}{\partial y} = 4y \qquad (1)$$

Integrate Eq. (1) with respect to y to obtain

$$\int d\psi = \int 4y\, dy$$

or

$$\psi = 2y^2 + f_1(x) \quad \text{where } f_1(x) \text{ is an arbitrary function of } x. \qquad (2)$$

Similarly,

$$v = -\frac{\partial \psi}{\partial x} = 4x$$

and

$$\int d\psi = -\int 4x\, dx$$

or

$$\psi = -2x^2 + f_2(y) \quad \text{where } f_2(y) \text{ is an arbitrary function of } y. \qquad (3)$$

To satisfy both Eqs. (2) and (3) $f_1(x) = f_2(y)$ for all x and y. Thus, $f_1 = f_2 = \text{constant}$.

$$\psi = \underline{\underline{2(y^2 - x^2) + C}}$$

Where C is a constant.

6.6R

6.6R (Velocity potential) A two-dimensional flow field is formed by adding a source at the origin of the coordinate system to the velocity potential

$$\phi = r^2 \cos 2\theta$$

Locate any stagnation points in the upper half of the coordinate plane ($0 \le \theta \le \pi$).

(ANS: $\theta_s = \pi/2$; $r_s = (m/4\pi)^{1/2}$)

$$\phi = \frac{m}{2\pi} \ln r + r^2 \cos 2\theta \text{ , where } \phi_{source} = \frac{m}{2\pi} \ln r$$

Thus,

$$v_\theta = \frac{1}{r}\frac{\partial \phi}{\partial \theta} = -2r \sin 2\theta$$

and

$$v_r = \frac{\partial \phi}{\partial r} = \frac{m}{2\pi r} + 2r \cos 2\theta$$

Stagnation points will occur where $v_\theta = 0$, $v_r = 0$, for $0 \le \theta \le \pi$.

Thus,

$$0 = -2r_s \sin 2\theta_s \qquad (1)$$

$$0 = \frac{m}{2\pi r_s} + 2r_s \cos 2\theta_s \qquad (2)$$

Equation (1) is satisfied at $r_s = 0$ or $\theta_s = 0, \frac{\pi}{2}, \pi$.

From Eq. (2)

$$\cos 2\theta_s = -\frac{m}{4\pi r_s^2} \qquad (3)$$

and for the possible values of θ_s, only $\theta_s = \frac{\pi}{2}$ will satisfy Eq. (3). Recall that $m > 0$ for a source.

Thus,

$$r_s = \sqrt{\frac{m}{4\pi}}$$

Thus, the stagnation point is located at

$$\underline{\underline{\theta_s = \frac{\pi}{2}}} \text{ , } \underline{\underline{r_s = \sqrt{\frac{m}{4\pi}}}}$$

6.7R

6.7R (Potential flow) The stream function for a two-dimensional, incompressible flow field is given by the equation

$$\psi = 2x - 2y$$

where the stream function has the units of ft^2/s with x and y in feet. **(a)** Sketch the streamlines for this flow field. Indicate the direction of flow along the streamlines. **(b)** Is this an irrotational flow field? **(c)** Determine the acceleration of a fluid particle at the point $x = 1$ ft, $y = 2$ ft.

(ANS: yes; no acceleration)

(a) Lines of constant ψ are streamlines. Thus, with $\psi = 2x - 2y$ the equation of a given streamline, ψ_1, (where ψ_1 is some constant) is of the form

$$\psi_1 = 2x - 2y$$

or

$$y = x - \frac{\psi_1}{2}$$

Thus, streamlines are straight lines as illustrated in the figure for three particular streamlines.

Since

$$u = \frac{\partial \psi}{\partial y} = -2 \qquad v = -\frac{\partial \psi}{\partial x} = -2$$

the direction of flow is as shown on the figure

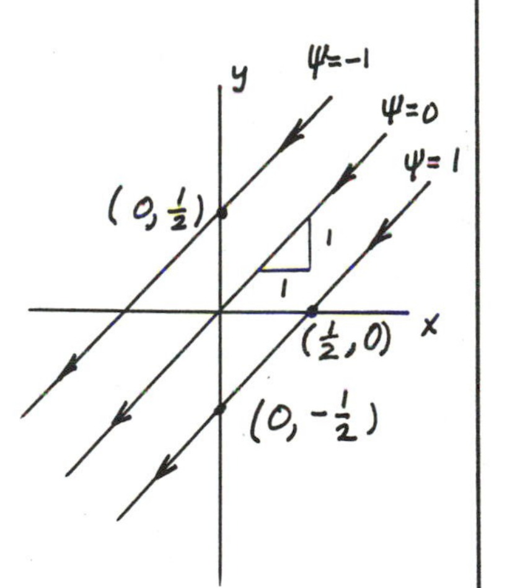

(b) The flow field is irrotational if $\omega_z = 0$ where

$$\omega_z = \frac{1}{2}\left(\frac{\partial v}{\partial x} - \frac{\partial u}{\partial y}\right) \qquad \text{(Eq. 6.12)}$$

For the stream function given

$$\frac{\partial v}{\partial x} = 0 \qquad \frac{\partial u}{\partial y} = 0$$

so that $\omega_z = 0$ and the flow field is irrotational. Yes.

(c) Since the velocity is constant throughout the flow field, the acceleration of all fluid particles is zero.

$$\vec{a} = \frac{\partial \vec{V}}{\partial t} + \vec{V} \cdot \nabla \vec{V} \equiv 0 \quad \text{since } \vec{V} = -2\hat{i} - 2\hat{j}$$

6.8R

6.8R (Inviscid flow) In a certain steady, incompressible, inviscid, two-dimensional flow field ($w = 0$, and all variables independent of z) the x component of velocity is given by the equation:

$$u = x^2 - y$$

Will the corresponding pressure gradient in the horizontal x direction be a function only of x, only of y, or of both x and y? Justify your answer.

(ANS: only of x)

Since the flow field must satisfy the continuity equation

$$\frac{\partial u}{\partial x} + \frac{\partial v}{\partial y} = 0$$

and with $u = x^2 - y$ it follows that

$$\frac{\partial v}{\partial y} = -\frac{\partial u}{\partial x} = -2x$$

and therefore

$$v = -2xy + f_1(x)$$

For steady, two-dimensional flow of an inviscid fluid (with the x-axis horizontal so that $g_x = 0$) the x-component of the momentum equation is

$$-\frac{\partial p}{\partial x} = \rho\left(u\frac{\partial u}{\partial x} + v\frac{\partial u}{\partial y}\right) \qquad \text{(Eq. 6.51a)}$$

Thus, for the u and v given above

$$\frac{\partial p}{\partial x} = -\rho\left[(x^2 - y)(2x) + (-2xy + f_1(x))(-1)\right]$$

$$= \rho\left[f_1(x) - 2x^3\right] = F(x)$$

The pressure gradient $\frac{\partial p}{\partial x}$ is a function only of x.

6.9R

6.9R (Inviscid flow) The stream function for the flow of a nonviscous, incompressible fluid in the vicinity of a corner (Fig. P6.9R) is

$$\psi = 2r^{4/3} \sin \tfrac{4}{3}\theta$$

Determine an expression for the pressure *gradient* along the boundary $\theta = 3\pi/4$.

(ANS: $-64\,\rho/27\,r^{1/3}$)

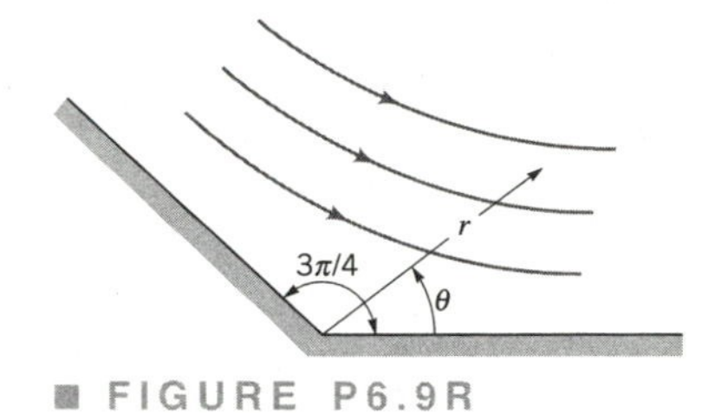

■ FIGURE P6.9R

Along the $\theta = 3\pi/4$ boundary, which is a streamline (i.e., $\psi = 0$ on $\theta = 3\pi/4$),

$$\frac{p}{\rho} + \frac{V^2}{2} = \text{constant}$$

or

$$\frac{\partial p}{\partial r} = -\rho V \frac{\partial V}{\partial r} \qquad (1)$$

For the stream function given,

$$v_r = \frac{1}{r}\frac{\partial \psi}{\partial \theta} = \frac{8}{3} r^{1/3} \cos \frac{4}{3}\theta$$

and along the $\theta = 3\pi/4$ boundary, $v_\theta = 0$, so that

$$V = v_r\left(\theta = \frac{3\pi}{4}\right) = -\frac{8}{3} r^{1/3}$$

Since

$$\frac{\partial V}{\partial r} = \frac{-8}{9} r^{-2/3}$$

it follows from Eq. (1) that

$$\frac{\partial p}{\partial r} = -\rho\left(-\frac{8}{3} r^{1/3}\right)\left(-\frac{8}{9} r^{-2/3}\right) = -\frac{64\,\rho}{27\,r^{1/3}}$$

6.10R (Potential flow) A certain body has the shape of a half-body with a thickness of 0.5 m. If this body is to be placed in an airstream moving at 20 m/s, what source strength is required to simulate flow around the body?
(ANS: 10.0 m^2/s)

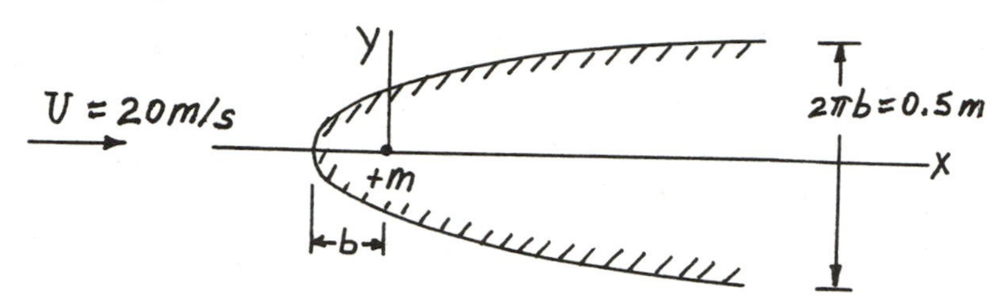

The width of half-body $= 2\pi b$ (see Fig. 6.24)
so that

$$b = \frac{(0.5\,m)}{2\pi}$$

From Eq. 6.99, the distance between the source and the nose of the body is

$$b = \frac{m}{2\pi U}$$

where m is the source strength, and therefore

$$m = 2\pi U b = 2\pi\left(20\,\frac{m}{s}\right)\left(\frac{0.5\,m}{2\pi}\right) = \underline{\underline{10.0\,\frac{m^2}{s}}}$$

6.11R

6.11R (Potential flow) A source and a sink are located along the x axis with the source at $x = -1$ ft and the sink at $x = 1$ ft. Both the source and the sink have a strength of 10 ft²/s. Determine the location of the stagnation points along the x axis when this source-sink pair is combined with a uniform velocity of 20 ft/s in the positive x direction.

(ANS: ±1.08 ft)

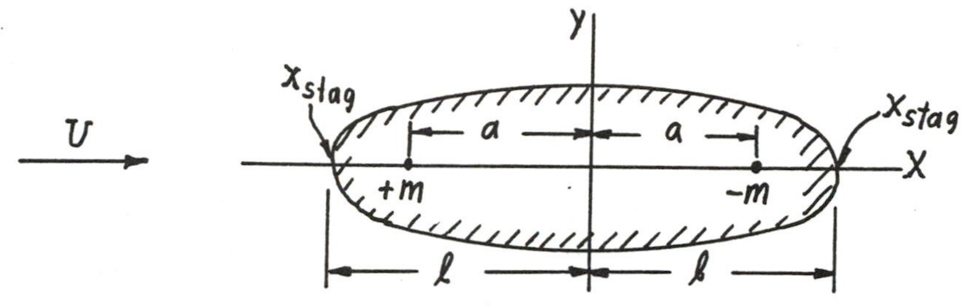

From Eq. 6.106

$$x_{stag} = \pm \ell = \pm \sqrt{\frac{ma}{\pi U} + a^2}$$

Thus, for $a = 1\ ft$, $m = 10\ \frac{ft^2}{s}$, and $U = 20\ \frac{ft}{s}$

$$x_{stag} = \pm \sqrt{\frac{(10\ \frac{ft^2}{s})(1\ ft)}{\pi\ (20\ \frac{ft}{s})} + (1\ ft)^2}$$

$$= \underline{\underline{\pm 1.08\ ft}}$$

6.12R

6.12R (Viscous flow) In a certain viscous, incompressible flow field with zero body forces the velocity components are

$$u = ay - b(cy - y^2)$$

$$v = w = 0$$

where a, b, and c are constant. **(a)** Use the Navier–Stokes equations to determine an expression for the pressure gradient in the x direction. **(b)** For what combination of the constants a, b, and c (if any) will the shearing stress, τ_{yx}, be zero at $y = 0$ where the velocity is zero?

(ANS: $2b\mu$; $a = bc$)

The x-component of the Navier-Stokes equation is (Eq. 6.127a)

(a) $$\rho\left(\frac{\partial u}{\partial t} + u\frac{\partial u}{\partial x} + v\frac{\partial u}{\partial y} + w\frac{\partial u}{\partial z}\right) = -\frac{\partial p}{\partial x} + \rho g_x + \mu\left(\frac{\partial^2 u}{\partial x^2} + \frac{\partial^2 u}{\partial y^2} + \frac{\partial^2 u}{\partial z^2}\right)$$

For the conditions specified, Eq. 6.127a above reduces to

$$0 = -\frac{\partial p}{\partial x} + 0 + \mu(2b)$$

so that

$$\underline{\underline{\frac{\partial p}{\partial x} = 2b\mu}}$$

(b) $$\tau_{yx} = \mu\left(\frac{\partial u}{\partial y} + \frac{\partial v}{\partial x}\right) \qquad \text{(Eq. 6.125d)}$$

For the given velocity distribution,

$$\frac{\partial u}{\partial y} = a - bc + 2by$$

and

$$\frac{\partial v}{\partial x} = 0$$

Thus,

$$\tau_{yx} = \mu(a - bc + 2by)$$

At $y = 0$

$$\tau_{yx} = \mu(a - bc)$$

and τ_{yx} will be zero if

$$\underline{\underline{a = bc}}$$

y

$u = by^2$

0 0 u

Note: With $v \equiv 0$ then $\tau_{yx} = \mu\frac{\partial u}{\partial y}$, where if $a = bc$, then $u = by^2$. Thus, $\frac{\partial u}{\partial y} = 2by = 0$ when $y = 0$.

6.13R

6.13R (Viscous flow) A viscous fluid is contained between two infinite, horizontal parallel plates that are spaced 0.5 in. apart. The bottom plate is fixed, and the upper plate moves with a constant velocity, U. The fluid motion is caused by the movement of the upper plate, and there is no pressure gradient in the direction of flow. The flow is laminar. If the velocity of the upper plate is 2 ft/s and the fluid has a viscosity of 0.03 lb·s/ft^2 and a specific weight of 70 lb/ft^3, what is the required horizontal force per square foot on the upper plate to maintain the 2 ft/s velocity? What is the pressure differential in the fluid between the top and bottom plates?

(ANS: 1.44 lb/ft^2; 2.92 lb/ft^2)

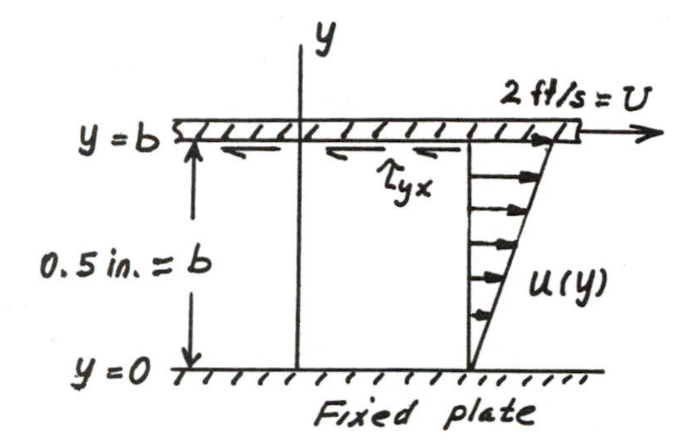

For simple Couette flow (see figure)

$$u = U\frac{y}{b} \qquad \text{(Eq. 6.142)}$$

Since,

$$\tau_{yx} = \mu\left(\frac{\partial u}{\partial y} + \frac{\partial v}{\partial x}\right) \qquad \text{(Eq. 6.125d)}$$

it follows that (with $v = 0$)

$$\tau_{yx} = \frac{\mu U}{b}$$

Thus,

$$\text{Force} = \tau_{yx} \times \text{area} = \frac{\left(0.03\ \frac{\text{lb}\cdot\text{s}}{\text{ft}^2}\right)\left(2\ \frac{\text{ft}}{\text{s}}\right)}{\left(\frac{0.5\ \text{in.}}{12\ \frac{\text{in.}}{\text{ft}}}\right)}\left(1\ \text{ft}^2\right)$$

$$= \underline{\underline{1.44\ \frac{\text{lb}}{\text{ft}^2}}}$$

Since there is no fluid motion in the y-direction, the pressure variation in the y-direction is hydrostatic so that

$$p_{\text{bottom}} - p_{\text{top}} = \gamma b = \left(70\ \frac{\text{lb}}{\text{ft}^3}\right)\left(\frac{0.5\ \text{in.}}{12\ \frac{\text{in.}}{\text{ft}}}\right) = \underline{\underline{2.92\ \frac{\text{lb}}{\text{ft}^2}}}$$

6.14R

6.14R (Viscous flow) A viscous liquid ($\mu = 0.016$ lb·s/ft^2, $\rho = 1.79$ slugs/ft^3) flows through the annular space between two horizontal, fixed, concentric cylinders. If the radius of the inner cylinder is 1.5 in. and the radius of the outer cylinder is 2.5 in., what is the volume flowrate when the pressure drop along the axis of the annulus is 100 lb/ft^2 per ft?

(ANS: 0.317 ft^3/s)

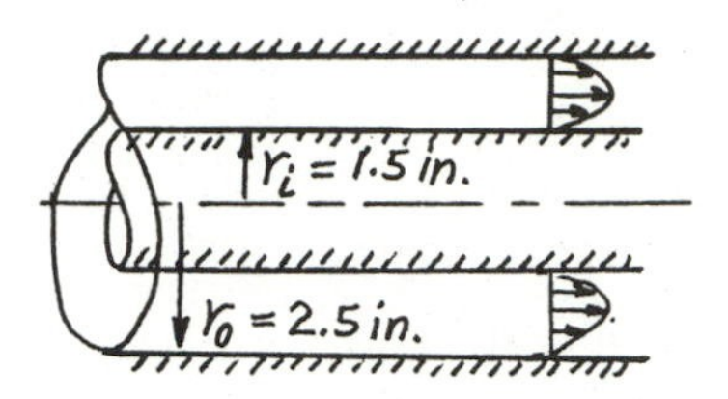

For laminar flow in an annulus (see Eq. 6.156):

$$Q = \frac{\pi}{8\mu}\frac{\Delta p}{\ell}\left[r_o^4 - r_i^4 - \frac{(r_o^2 - r_i^2)^2}{\ln(r_o/r_i)}\right] \quad \text{where } \frac{\Delta p}{\ell} = 100\,\frac{lb}{ft^3}$$

so that

$$Q = \frac{\pi}{8\left(0.016\,\frac{lb\cdot s}{ft^2}\right)}\left(100\,\frac{lb}{ft^3}\right)\left[\left(\frac{2.5}{12}ft\right)^4 - \left(\frac{1.5}{12}ft\right)^4 - \frac{\left[\left(\frac{2.5}{12}ft\right)^2 - \left(\frac{1.5}{12}ft\right)^2\right]^2}{\ln\left(\frac{2.5\,in.}{1.5\,in.}\right)}\right]$$

$$= 0.317\,\frac{ft^3}{s}$$

We must check to see if the flow is laminar. If not, the above result is not correct.

Now,

$$Re = \frac{\rho V D_h}{\mu}, \quad \text{where } D_h = 2(r_o - r_i)$$

or

$$D_h = 2\left(\frac{2.5\,in. - 1.5\,in.}{12\,in./ft}\right) = 0.167\,ft$$

Also,

$$V = \frac{Q}{A} = \frac{Q}{\pi(r_o^2 - r_i^2)} = \frac{0.317\,\frac{ft^3}{s}}{\pi(2.5^2 - 1.5^2)\,in.^2}\,\frac{144\,in.^2}{ft^2} = 3.63\,\frac{ft}{s}$$

Thus,

$$Re = \frac{\left(1.79\,\frac{slugs}{ft^3}\right)\left(3.63\,\frac{ft}{s}\right)(0.167\,ft)}{0.016\,\frac{lb\cdot s}{ft^2}} = 67.8$$

Since $Re = 67.8 \ll 2100$ the flow is laminar and

$$\underline{\underline{Q = 0.317\,\frac{ft^3}{s}}}$$

6.15R

6.15R (Viscous flow) Consider the steady, laminar flow of an incompressible fluid through the horizontal rectangular channel of Fig. P6.15R. Assume that the velocity components in the x and y directions are zero and the only body force is the weight. Start with the Navier–Stokes equations. **(a)** Determine the appropriate set of differential equations and boundary conditions for this problem. You need not solve the equations. **(b)** Show that the pressure distribution is hydrostatic at any particular cross section.

(ANS: $\partial p/\partial x = 0$; $\partial p/\partial y = -\rho g$; $\partial p/\partial z = \mu(\partial^2 w/\partial x^2 + \partial^2 w/\partial y^2)$ with $w = 0$ for $x = \pm b/2$ and $y = \pm a/2$)

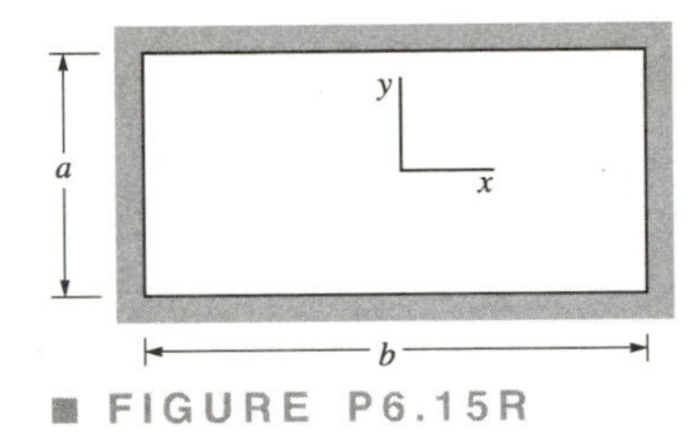

■ FIGURE P6.15R

(a) From the description of the problem, $u = 0$, $v = 0$, $g_x = 0$, $g_y = -g$, $g_z = 0$, $w \neq f(t)$, and the continuity equation indicates that $\frac{\partial w}{\partial z} = 0$. With these conditions the Navier-Stokes equations (Eq. 6.127) reduce to

(x-direction) $$\frac{\partial p}{\partial x} = 0 \quad (1)$$

(y-direction) $$\frac{\partial p}{\partial y} = -\rho g \quad (2)$$

(z-direction) $$\frac{\partial p}{\partial z} = \mu\left(\frac{\partial^2 w}{\partial x^2} + \frac{\partial^2 w}{\partial y^2}\right) \quad (3)$$

Boundary conditions are: $w = 0$ for $x = \pm \frac{b}{2}$, $y = \pm \frac{a}{2}$

(b) Integration of Eq. (2) yields

$$\int dp = -\int \rho g \, dy$$

$$p = -\rho g y + f_1(x, z) \quad (4)$$

However, from Eq. (1), p is not a function of x so that Eq. (4) becomes

$$p = -\rho g y + f_1(z)$$

Thus, at a given cross section (z = constant)

$$p = -\rho g y + C \quad (5)$$

where C is a constant. Equation (5) indicates that the pressure distribution is hydrostatic at a given cross section.

6.16R

6.16R (Viscous flow) A viscous liquid, having a viscosity of 10^{-4} lb·s/ft² and a specific weight of 50 lb/ft³, flows steadily through the 2-in.-diameter, horizontal, smooth pipe shown in Fig. P6.16R. The mean velocity in the pipe is 0.5 ft/s. Determine the differential reading, Δh, on the inclined-tube manometer.

(ANS: 0.0640 ft)

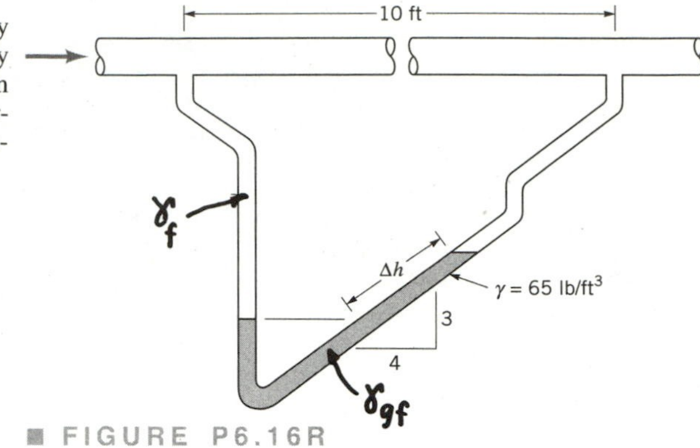

■ FIGURE P6.16R

Check Reynolds number to determine if flow is laminar:

$$Re = \frac{\rho V (2R)}{\mu} = \frac{\left(\frac{50\,\frac{lb}{ft^3}}{32.2\,ft/s^2}\right)\left(0.5\,\frac{ft}{s}\right)\left(\frac{2\,in.}{12\,in./ft}\right)}{10^{-4}\,\frac{lb\cdot s}{ft^2}} = 1290 < 2100$$

Thus, flow is laminar and

$$V = \frac{R^2}{8\mu}\frac{\Delta p}{\ell} \qquad \text{(Eq. 6.152)}$$

so that

$$\Delta p = \frac{8\mu \ell V}{R^2} = \frac{8\left(10^{-4}\,\frac{lb\cdot s}{ft^2}\right)(10\,ft)\left(0.5\,\frac{ft}{s}\right)}{\left(\frac{1\,in.}{12\,\frac{in.}{ft}}\right)^2} = 0.576\,\frac{lb}{ft^2}$$

For manometer (see figure)

$$p_1 + \gamma_f\left(\tfrac{3}{5}\Delta h\right) - \gamma_{gf}\left(\tfrac{3}{5}\Delta h\right) = p_2$$

or

$$p_1 - p_2 = \Delta p = \tfrac{3}{5}\Delta h\left(\gamma_{gf} - \gamma_f\right)$$

Thus,

$$\Delta h = \frac{\frac{5}{3}\Delta p}{\gamma_{gf} - \gamma_f} = \frac{\frac{5}{3}\left(0.576\,\frac{lb}{ft^2}\right)}{65\,\frac{lb}{ft^3} - 50\,\frac{lb}{ft^3}} = \underline{\underline{0.0640\ ft}}$$

7

Similitude, Dimensional Analysis, and Modeling

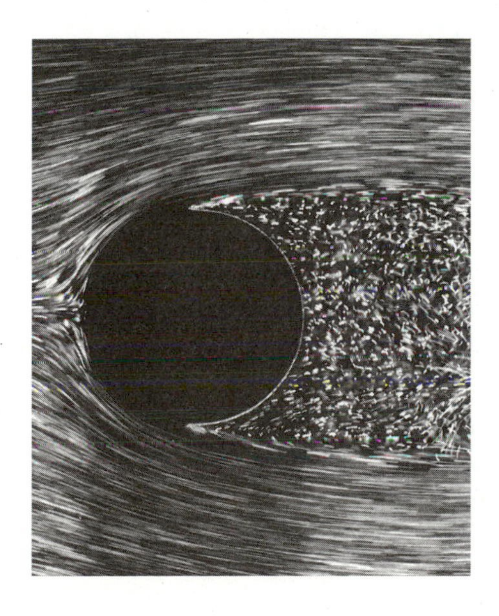

Flow past a circular cylinder with Re = 2000: The streaklines of flow past any circular cylinder (regardless of size, velocity, or fluid) are as shown provided that the dimensionless parameter called the Reynolds number, Re, is equal to 2000. For other values of Re the flow pattern will be different (air bubbles in water). (Photograph courtesy of ONERA, France.)

7.1R

7.1R (Common Pi terms) Standard air with velocity V flows past an airfoil having a chord length, b, of 6 ft. **(a)** Determine the Reynolds number, $\rho V b/\mu$, for $V = 150$ mph. **(b)** If this airfoil were attached to an airplane flying at the same speed in a standard atmosphere at an altitude of 10,000 ft, what would be the value of the Reynolds number?

(ANS: 8.40×10^6; 6.56×10^6)

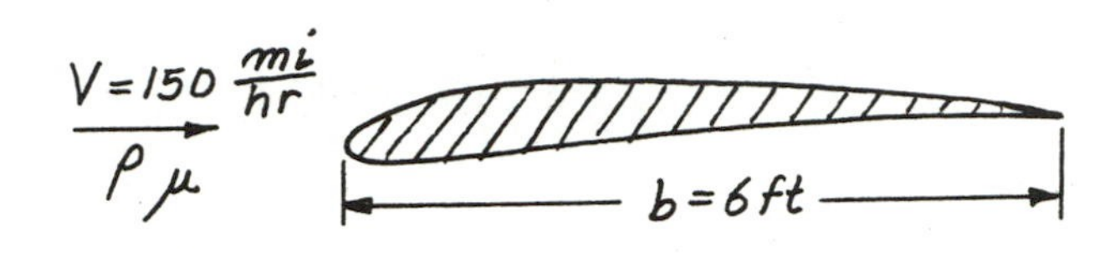

(a) $$Re = \frac{\rho V b}{\mu} = \frac{\left(2.38\times10^{-3}\ \frac{\text{slugs}}{\text{ft}^3}\right)\left(150\ \frac{\text{mi}}{\text{hr}}\times 5280\ \frac{\text{ft}}{\text{mi}}\times\frac{1}{3600\ \frac{\text{s}}{\text{hr}}}\right)(6\ \text{ft})}{3.74\times10^{-7}\ \frac{\text{lb}\cdot\text{s}}{\text{ft}^2}}$$

or

$$Re = \underline{\underline{8.40\times10^6}}$$

(b) At 10,000 ft, $\rho = 1.756\times10^{-3}\ \frac{\text{slugs}}{\text{ft}^3}$ and $\mu = 3.534\times10^{-7}\ \frac{\text{lb}\cdot\text{s}}{\text{ft}^2}$ (Table C.1 in Appendix C) so that

$$\frac{\rho V b}{\mu} = \frac{\left(1.756\times10^{-3}\ \frac{\text{slugs}}{\text{ft}^3}\right)\left(150\ \frac{\text{mi}}{\text{hr}}\times 5280\ \frac{\text{ft}}{\text{mi}}\times\frac{1}{3600\ \frac{\text{s}}{\text{hr}}}\right)(6\ \text{ft})}{3.534\times10^{-7}\ \frac{\text{lb}\cdot\text{s}}{\text{ft}^2}}$$

or

$$Re = \underline{\underline{6.56\times10^6}}$$

7.2R

7.2R (Dimensionless variables) Some common variables in fluid mechanics include: volume flowrate, Q, acceleration of gravity, g, viscosity, μ, density, ρ and a length, ℓ. Which of the following combinations of these variables are dimensionless? **(a)** $Q^2/g\ell^2$. **(b)** $\rho Q/\mu\ell$. **(c)** $g\ell^5/Q^2$. **(d)** $\rho Q\ell/\mu$.

(ANS: (b); (c))

$$Q \doteq L^3T^{-1} \qquad g \doteq LT^{-2} \qquad \mu \doteq FL^{-2}T \qquad \rho \doteq FL^{-4}T^2 \qquad \ell \doteq L$$

(a) $$\frac{Q^2}{g\ell^2} \doteq \frac{(L^3T^{-1})^2}{(LT^{-2})(L^2)} \doteq L^3$$ not dimensionless

(b) $$\frac{\rho Q}{\mu \ell} \doteq \frac{(FL^{-4}T^2)(L^3T^{-1})}{(FL^{-2}T)(L)} \doteq F^0L^0T^0$$ dimensionless

(c) $$\frac{g\ell^5}{Q^2} \doteq \frac{(LT^{-2})(L^5)}{(L^3T^{-1})^2} \doteq L^0T^0$$ dimensionless

(d) $$\frac{\rho Q\ell}{\mu} \doteq \frac{(FL^{-4}T^2)(L^3T^{-1})(L)}{FL^{-2}T} \doteq L^2$$ not dimensionless

7.3R

7.3R (Determination of Pi terms) A fluid flows at a velocity V through a horizontal pipe of diameter D. An orifice plate containing a hole of diameter d is placed in the pipe. It is desired to investigate the pressure drop, Δp, across the plate. Assume that

$$\Delta p = f(D, d, \rho, V)$$

where ρ is the fluid density. Determine a suitable set of pi terms. (ANS: $\Delta p/\rho V^2 = \phi(d/D)$)

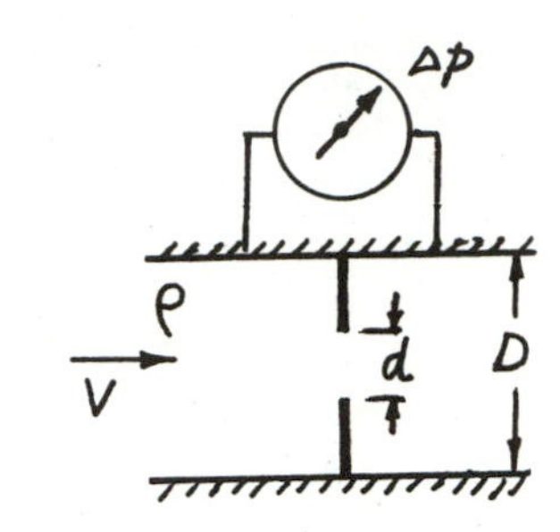

$\Delta p \doteq FL^{-2} \quad D \doteq L \quad d \doteq L \quad \rho \doteq FL^{-4}T^{2} \quad V \doteq LT^{-1}$

From the pi theorem, $5-3=2$ pi terms required. Use D, V, and ρ as repeating variables. Thus,

$$\Pi_1 = \Delta p\, D^a V^b \rho^c$$

and

$$(FL^{-2})(L)^a(LT^{-1})^b(FL^{-4}T^{2})^c \doteq F^0L^0T^0$$

so that

$$1 + c = 0 \qquad \text{(for } F\text{)}$$
$$-2 + a + b - 4c = 0 \qquad \text{(for } L\text{)}$$
$$-b + 2c = 0 \qquad \text{(for } T\text{)}$$

It follows that $a = 0$, $b = -2$, $c = -1$, and therefore

$$\Pi_1 = \frac{\Delta p}{V^2 \rho}$$

Check dimensions using MLT system:

$$\frac{\Delta p}{V^2\rho} \doteq \frac{ML^{-1}T^{-2}}{(LT^{-1})^2(ML^{-3})} \doteq M^0L^0T^0 \qquad \therefore \text{OK}$$

For Π_2:

$$\Pi_2 = d\, D^a V^b \rho^c$$

$$(L)(L)^a(LT^{-1})^b(FL^{-4}T^{2})^c \doteq F^0L^0T^0$$

$$c = 0 \qquad \text{(for } F\text{)}$$
$$1 + a + b - 4c = 0 \qquad \text{(for } L\text{)}$$
$$-b + 2c = 0 \qquad \text{(for } T\text{)}$$

It follows that $a = -1$, $b = 0$, $c = 0$, and therefore

$$\Pi_2 = \frac{d}{D}$$

which is obviously dimensionless.

Thus,

$$\frac{\Delta p}{\rho V^2} = \phi\left(\frac{d}{D}\right)$$

7.4R

7.4R (Determination of Pi terms) The flowrate, Q, in an open canal or channel can be measured by placing a plate with a V-notch across the channel as illustrated in Fig. P7.4R. This type of device is called a V-notch *weir*. The height, H, of the liquid above the crest can be used to determine Q. Assume that

$$Q = f(H, g, \theta)$$

where g is the acceleration of gravity. What are the significant dimensionless parameters for this problem?

(ANS: $Q/(gH^5)^{1/2} = \phi(\theta)$)

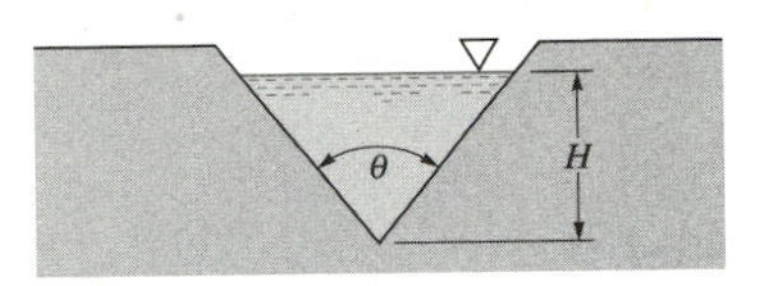

■ FIGURE P7.4R

$$Q \doteq L^3T^{-1} \qquad H \doteq L \qquad g \doteq LT^{-2} \qquad \theta = F^0L^0T^0$$

From the pi theorem, $4-2 = 2$ pi terms required: $\Pi_1 = \phi(\Pi_2)$

By inspection, for Π_1 (containing Q):

$$\Pi_1 = \frac{Q}{g^{1/2}H^{5/2}} \doteq \frac{L^3T^{-1}}{(LT^{-2})^{1/2}(L)^{5/2}} \doteq L^0T^0$$

Since the angle, θ, is dimensionless

$$\Pi_2 = \theta$$

So that

$$\frac{Q}{\sqrt{gH^5}} = \phi(\theta)$$

7.5R (Determination of Pi terms) In a fuel injection system, small droplets are formed due to the breakup of the liquid jet. Assume the droplet diameter, d, is a function of the liquid density, ρ, viscosity, μ, and surface tension, σ, and the jet velocity, V, and diameter, D. Form an appropriate set of dimensionless parameters using μ, V, and D as repeating variables.
(ANS: $d/D = \phi(\rho VD/\mu,\ \sigma/\mu V)$)

$$d = f(\rho, \mu, \sigma, V, D)$$

$d \doteq L \quad \rho \doteq FL^{-4}T^{2} \quad \mu \doteq FL^{-2}T \quad \sigma \doteq FL^{-1} \quad V \doteq LT^{-1} \quad D \doteq L$

From the pi theorem, 6-3 = 3 pi terms required. Use μ, V, and D as repeating variables. Thus,

$$\Pi_1 = d\,\mu^{a}V^{b}D^{c}$$

and

$$(L)(FL^{-2}T)^{a}(LT^{-1})^{b}(L)^{c} \doteq F^{0}L^{0}T^{0}$$

so that

$$a = 0 \qquad \text{(for } F\text{)}$$
$$1 - 2a + b + c = 0 \qquad \text{(for } L\text{)}$$
$$a - b = 0 \qquad \text{(for } T\text{)}$$

It follows that $a = 0$, $b = 0$, $c = -1$, and therefore

$$\Pi_1 = \frac{d}{D}$$

which is obviously dimensionless.

For Π_2:

$$\Pi_2 = \rho\,\mu^{a}V^{b}D^{c}$$

$$(FL^{-4}T^{2})(FL^{-2}T)^{a}(LT^{-1})^{b}(L)^{c} \doteq F^{0}L^{0}T^{0}$$

$$1 + a = 0 \qquad \text{(for } F\text{)}$$
$$-4 - 2a + b + c = 0 \qquad \text{(for } L\text{)}$$
$$2 + a - b = 0 \qquad \text{(for } T\text{)}$$

It follows that $a = -1$, $b = 1$, $c = 1$, and therefore

$$\Pi_2 = \frac{\rho V D}{\mu}$$

(continued)

Check dimensions using MLT system:

$$\frac{\rho V D}{\mu} \doteq \frac{(ML^{-3})(LT^{-1})(L)}{ML^{-1}T^{-1}} \doteq M^0L^0T^0 \quad \therefore \text{OK}$$

For Π_3:

$$\Pi_3 = \sigma \mu^a V^b D^c$$

$$(FL^{-1})(FL^{-2}T)^a (LT^{-1})^b (L)^c \doteq F^0L^0T^0$$

$$1 + a = 0 \qquad \text{(for } F\text{)}$$

$$-1 - 2a + b + c = 0 \qquad \text{(for } L\text{)}$$

$$a - b = 0 \qquad \text{(for } T\text{)}$$

It follows that $a = -1$, $b = -1$, $c = 0$, and therefore

$$\Pi_3 = \frac{\sigma}{\mu V}$$

Check dimensions using MLT system:

$$\frac{\sigma}{\mu V} \doteq \frac{MT^{-2}}{(ML^{-1}T^{-1})(LT^{-1})} \doteq M^0L^0T^0 \quad \therefore \text{OK}$$

Thus,

$$\frac{d}{D} = \phi\left(\frac{\rho V D}{\mu}, \frac{\sigma}{\mu V}\right)$$

7.6R

7.6R (Determination of Pi terms) The thrust, $\mathcal{T}$, developed by a propeller of a given shape depends on its diameter, D, the fluid density, ρ, and viscosity, μ, the angular speed of rotation, ω, and the advance velocity, V. Develop a suitable set of pi terms, one of which should be $\rho D^2 \omega / \mu$. Form the pi terms by inspection.

(ANS: $\mathcal{T}/\rho V^2 D^2 = \phi(\rho V D/\mu,\ \rho D^2 \omega/\mu)$)

$$\mathcal{T} = f(D, \rho, \mu, \omega, V)$$

$$\mathcal{T} \doteq F \quad D \doteq L \quad \rho \doteq FL^{-4}T^2 \quad \mu \doteq FL^{-2}T \quad \omega \doteq T^{-1} \quad V \doteq LT^{-1}$$

From the pi theorem, $6-3=3$ pi terms required: $\Pi_1 = \phi(\Pi_2, \Pi_3)$.
By inspection, for Π_1 (containing $\mathcal{T}$):

$$\Pi_1 = \frac{\mathcal{T}}{\rho V^2 D^2} \doteq \frac{F}{(FL^{-4}T^2)(LT^{-1})^2(L)^2} \doteq F^0L^0T^0$$

Check using MLT:

$$\frac{\mathcal{T}}{\rho V^2 D^2} \doteq \frac{MLT^{-2}}{(ML^{-3})(LT^{-1})^2(L)^2} \doteq M^0L^0T^0 \quad \therefore \text{ok}$$

For Π_2 (containing μ):

$$\Pi_2 = \frac{\rho V D}{\mu}$$

which is the Reynolds number (known to be dimensionless).

For Π_3 (containing ω):

$$\Pi_3 = \frac{\omega \rho D^2}{\mu} \doteq \frac{(T^{-1})(FL^{-4}T^2)(L)^2}{FL^{-2}T} \doteq F^0L^0T^0$$

Check using MLT:

$$\frac{\omega \rho D^2}{\mu} \doteq \frac{(T^{-1})(ML^{-3})(L)^2}{ML^{-1}T^{-1}} \doteq M^0L^0T^0 \quad \therefore \text{ok}$$

Thus,

$$\frac{\mathcal{T}}{\rho V^2 D^2} = \phi\left(\frac{\rho V D}{\mu}, \frac{\rho D^2 \omega}{\mu}\right)$$

7.7R

7.7R (Modeling/similarity) The water velocity at a certain point along a 1:10 scale model of a dam spillway is 5 m/s. What is the corresponding prototype velocity if the model and prototype operate in accordance with Froude number similarity?

(ANS: 15.8 m/s)

$V_m = 5\,\frac{m}{s}$

$\ell_m = \frac{1}{10}\ell$

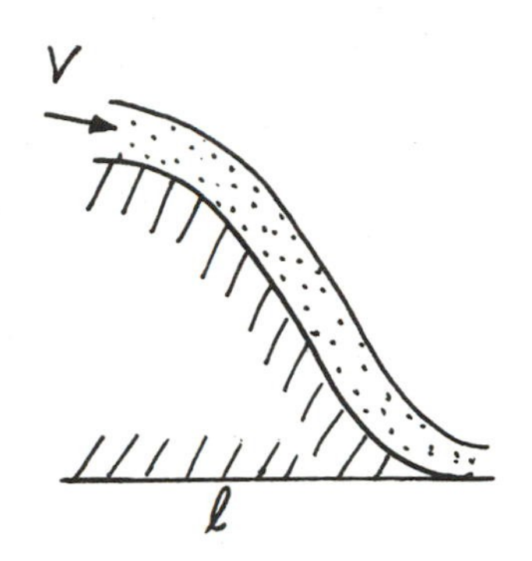

For Froude number similarity, $Fr_m = Fr$, or

$$\frac{V_m}{\sqrt{g_m \ell_m}} = \frac{V}{\sqrt{g \ell}}$$

so that

$$V = \sqrt{\left(\frac{g}{g_m}\right)\left(\frac{\ell}{\ell_m}\right)}\; V_m$$

and with $g = g_m$, $\ell/\ell_m = 10$, $V_m = 5\ m/s$, then

$$V = \sqrt{10}\left(5\,\frac{m}{s}\right) = \underline{\underline{15.8\ \frac{m}{s}}}$$

7.8R

7.8R (Modeling/similarity) The pressure drop per unit length in a 0.25-in.-diameter gasoline fuel line is to be determined from a laboratory test using the same tubing but with water as the fluid. The pressure drop at a gasoline velocity of 1.0 ft/s is of interest. **(a)** What water velocity is required? **(b)** At the properly scaled velocity from part (a), the pressure drop per unit length (using water) was found to be 0.45 psf/ft. What is the predicted pressure drop per unit length for the gasoline line?

(ANS: 2.45 ft/s; 0.0510 lb/ft^2 per ft)

For flow in a closed conduit,

$$\text{Dependent pi term} = \phi\left(\frac{\ell_i}{\ell}, \frac{\varepsilon}{\ell}, \frac{\rho V \ell}{\mu}\right) \qquad (\text{Eq. } 7.16)$$

For this particular problem the dependent variable is the pressure drop per unit length, Δp_ℓ, so that

$$\text{Dependent pi term} = \frac{\Delta p_\ell\, \ell}{\rho V^2} \doteq \frac{(FL^{-2})(L)}{(FL^{-4}T^2)(L^2T^{-2})} = F^0L^0T^0$$

Also, the characteristic length for pipe flow is the pipe diameter, D. Thus,

$$\frac{\Delta p_\ell\, D}{\rho V^2} = \phi\left(\frac{\varepsilon}{D}, \frac{\rho V D}{\mu}\right) \qquad (1)$$

(a) To maintain dynamic similarity,

$$\frac{\rho_m V_m D_m}{\mu_m} = \frac{\rho V D}{\mu}$$

and with $D_m = D$

$$V_m = \frac{\mu_m}{\mu}\frac{\rho}{\rho_m} V$$

$$= \frac{\left(2.34 \times 10^{-5}\ \frac{lb \cdot s}{ft^2}\right)}{\left(6.5 \times 10^{-6}\ \frac{lb \cdot s}{ft^2}\right)} \frac{\left(1.32\ \frac{slugs}{ft^3}\right)}{\left(1.94\ \frac{slugs}{ft^3}\right)} \left(1\ \frac{ft}{s}\right) = \underline{\underline{2.45\ \frac{ft}{s}}}$$

(b) With the same Reynolds number, and with $\varepsilon_m / D_m = \varepsilon / D$ (same tubing), then from Eq. (1)

$$\frac{\Delta p_{\ell m}\, D_m}{\rho_m V_m^2} = \frac{\Delta p_\ell\, D}{\rho V^2}$$

so that

$$\Delta p_\ell = \frac{\rho}{\rho_m}\frac{V^2}{V_m^2}\frac{D_m}{D}\Delta p_{\ell m} = \frac{\left(1.32\ \frac{slugs}{ft^3}\right)}{\left(1.94\ \frac{slugs}{ft^3}\right)} \frac{\left(1\ \frac{ft}{s}\right)^2}{\left(2.45\ \frac{ft}{s}\right)^2} (1)\left(0.45\ \frac{psf}{ft}\right) = \underline{\underline{0.0510\ \frac{psf}{ft}}}$$

7.9R

7.9R (Modeling/similarity) A thin layer of an incompressible fluid flows steadily over a horizontal smooth plate as shown in Fig. P7.9R. The fluid surface is open to the atmosphere, and an obstruction having a square cross section is placed on the plate as shown. A model with a length scale of $\frac{1}{4}$ and a fluid density scale of 1.0 is to be designed to predict the depth of fluid, y, along the plate. Assume that inertial, gravitational, surface tension, and viscous effects are all important. What are the required viscosity and surface tension scales?

(ANS: 0.125; 0.0625)

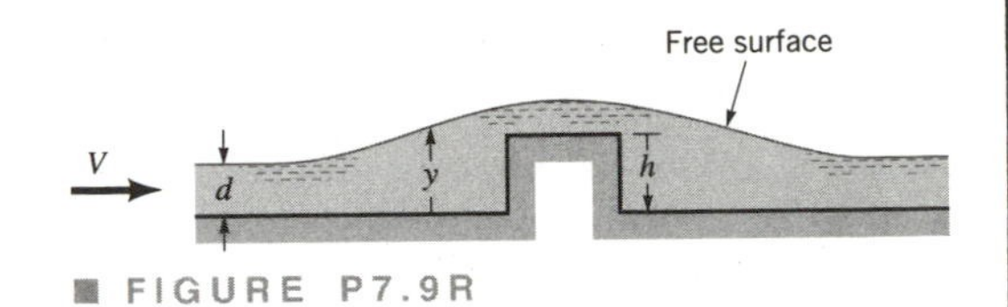

■ FIGURE P7.9R

A fluid dynamics problem for which inertial, gravitational, surface tension, and viscous effects are all important requires Froude, Reynolds, and Weber number similarity (see Table 7.1). Thus, for

$$Fr_m = Fr \quad \text{or} \quad \frac{V_m}{\sqrt{g_m d_m}} = \frac{V}{\sqrt{g d}}$$

(Froude number similarity) it follows that (with $g = g_m$)

$$\frac{V_m}{V} = \sqrt{\frac{d_m}{d}}$$

For Reynolds number similarity,

$$Re_m = Re \quad \text{or} \quad \frac{\rho_m V_m d_m}{\mu_m} = \frac{\rho V d}{\mu}$$

so that

$$\frac{\mu_m}{\mu} = \frac{\rho_m}{\rho}\frac{V_m}{V}\frac{d_m}{d} = \frac{\rho_m}{\rho}\sqrt{\frac{d_m}{d}}\,\frac{d_m}{d} = \frac{\rho_m}{\rho}\left(\frac{d_m}{d}\right)^{3/2}$$

$$= (1.0)\left(\frac{1}{4}\right)^{3/2} = \frac{1}{8} = \underline{\underline{0.125}}$$

For Weber number similarity,

$$We_m = We \quad \text{or} \quad \frac{\rho_m V_m^2 d_m}{\sigma_m} = \frac{\rho V^2 d}{\sigma}$$

Hence,

$$\frac{\sigma_m}{\sigma} = \frac{\rho_m}{\rho}\frac{V_m^2}{V^2}\frac{d_m}{d} = \frac{\rho_m}{\rho}\left(\sqrt{\frac{d_m}{d}}\right)^2\frac{d_m}{d} = \frac{\rho_m}{\rho}\left(\frac{d_m}{d}\right)^2$$

$$= (1.0)\left(\frac{1}{4}\right)^2 = \frac{1}{16} = \underline{\underline{0.0625}}$$

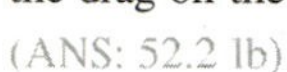
7.10R

7.10R (Correlation of experimental data) The drag on a 30-ft long, vertical, 1.25-ft diameter pole subjected to a 30 mph wind is to be determined with a model study. It is expected that the drag is a function of the pole length and diameter, the fluid density and viscosity, and the fluid velocity. Laboratory model tests were performed in a high-speed water tunnel using a model pole having a length of 2 ft and a diameter of 1 in. Some model drag data are shown in Fig. P7.10R. Based on these data, predict the drag on the full-sized pole.

(ANS: 52.2 lb)

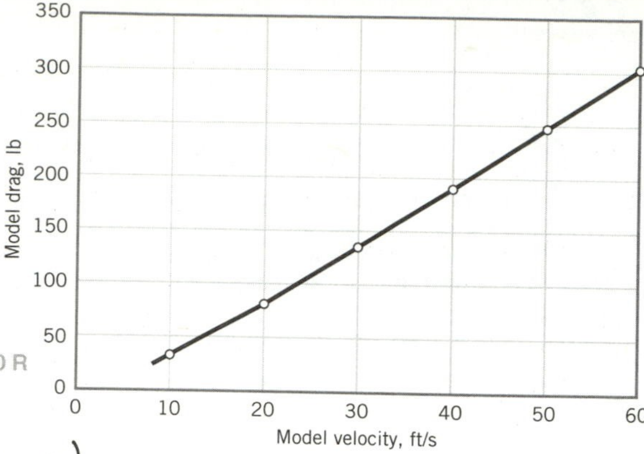

■ FIGURE P7.10R

$$\mathcal{D} = f(\ell, D, \rho, \mu, V)$$

where: $\mathcal{D} \sim$ drag $\doteq F$, $\ell \sim$ pole length $\doteq L$, $D \sim$ pole diameter $\doteq L$, $\rho \sim$ fluid density $\doteq FL^{-4}T^2$, $\mu \sim$ fluid viscosity $\doteq FL^{-2}T$, $V \sim$ velocity $\doteq LT^{-1}$.

From the pi theorem, 6-3 = 3 pi terms required. A dimensional analysis (for example using D, V, ρ, as repeating variables yields

$$\frac{\mathcal{D}}{\rho V^2 D^2} = \phi\left(\frac{\ell}{D}, \frac{\rho V D}{\mu}\right)$$

Geometric similarity requires that $\frac{\ell_m}{D_m} = \frac{\ell}{D}$ and this condition is satisfied since

$$\frac{\ell}{D} = \frac{2\,ft}{1/12\,ft} = \frac{30\,ft}{1.25\,ft} = 24 = \frac{\ell_m}{D_m}$$

Reynolds number similarity requires that $\frac{\rho_m V_m D_m}{\mu_m} = \frac{\rho V D}{\mu}$

so that

$$V_m = \frac{\mu_m}{\mu}\frac{\rho}{\rho_m}\frac{D}{D_m}V = \left(\frac{2.34\times10^{-5}\,\frac{lb\cdot s}{ft^2}}{3.74\times10^{-7}\,\frac{lb\cdot s}{ft^2}}\right)\left(\frac{0.00238\,\frac{slugs}{ft^3}}{1.94\,\frac{slugs}{ft^3}}\right)\left(\frac{1.25\,ft}{1/12\,ft}\right)V$$

$$= 1.15\,V$$

and with

$$V = (30\,mph)\left(\frac{1\,hr}{3600\,s}\right)\left(\frac{5280\,ft}{mi}\right) = 44.0\,\frac{ft}{s}$$

it follows that the required model velocity is $V_m = 1.15\left(44.0\,\frac{ft}{s}\right) = 50.6\,\frac{ft}{s}$

From the figure, at $V_m = 50.6\,\frac{ft}{s}$ the model drag $\mathcal{D}_m = 250$ lb. The prediction equation is

$$\frac{\mathcal{D}}{\rho V^2 D^2} = \frac{\mathcal{D}_m}{\rho_m V_m^2 D_m^2}$$

and therefore

$$\mathcal{D} = \frac{\rho}{\rho_m}\frac{V^2}{V_m^2}\frac{D^2}{D_m^2}\mathcal{D}_m = \left(\frac{0.00238\,\frac{slugs}{ft^3}}{1.94\,\frac{slugs}{ft^3}}\right)\left(\frac{44.0\,\frac{ft}{s}}{50.6\,\frac{ft}{s}}\right)^2\left(\frac{1.25\,ft}{1/12\,ft}\right)^2(250\,lb)$$

Thus, the predicted drag is

$$\mathcal{D} = \underline{\underline{52.2\,lb}}$$

7.11R

7.11R (Correlation of experimental data) A liquid is contained in a U-tube as is shown in Fig. P7.11R. When the liquid is displaced from its equilibrium position and released, it oscillates with a period τ. Assume that τ is a function of the acceleration of gravity, g, and the column length, ℓ. Some laboratory measurements made by varying ℓ and measuring τ, with $g = 32.2 \text{ ft/s}^2$, are given in the following table.

τ (s)	0.548	0.783	0.939	1.174
ℓ (ft)	0.49	1.00	1.44	2.25

Based on these data, determine a general equation for the period. (ANS: $\tau = 4.44(\ell/g)^{1/2}$)

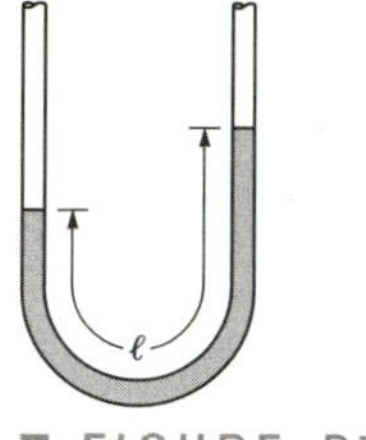

■ FIGURE P7.11R

$$\tau = f(g, \ell)$$

$$\tau \doteq T \qquad g \doteq LT^{-2} \qquad \ell \doteq L$$

From the pi theorem, 3-2 = 1 pi term required: Π_1 = constant.

By inspection:

$$\Pi_1 = \tau\sqrt{\frac{g}{\ell}} \doteq \frac{T\,(LT^{-2})^{1/2}}{(L)^{1/2}} \doteq L^0T^0$$

Since there is only 1 pi term, it follows that

$$\tau\sqrt{\frac{g}{\ell}} = C$$

where C is a constant. For the data given:

ℓ, ft	0.49	1.00	1.44	2.25
$\tau\sqrt{\frac{g}{\ell}}$	4.44	4.44	4.44	4.44

Thus, C = 4.44 and

$$\underline{\underline{\tau = 4.44\sqrt{\frac{\ell}{g}}}}$$

7.12R

7.12R (Dimensionless governing equations) An incompressible fluid is contained between two large parallel plates as shown in Fig. P7.12R. The upper plate is fixed. If the fluid is initially at rest and the bottom plate suddenly starts to move with a constant velocity, U, the governing differential equation describing the fluid motion is

$$\rho \frac{\partial u}{\partial t} = \mu \frac{\partial^2 u}{\partial y^2}$$

where u is the velocity in the x direction, and ρ and μ are the fluid density and viscosity, respectively. Rewrite the equation and the initial and boundary conditions in dimensionless form using h and U as reference parameters for length and velocity, and $h^2\rho/\mu$ as a reference parameter for time.

(ANS: $\partial u^*/\partial t^* = \partial^2 u^*/\partial y^{*2}$ with $u^* = 0$ at $t^* = 0$, $u^* = 1$ at $y^* = 0$, and $u^* = 0$ at $y^* = 1$)

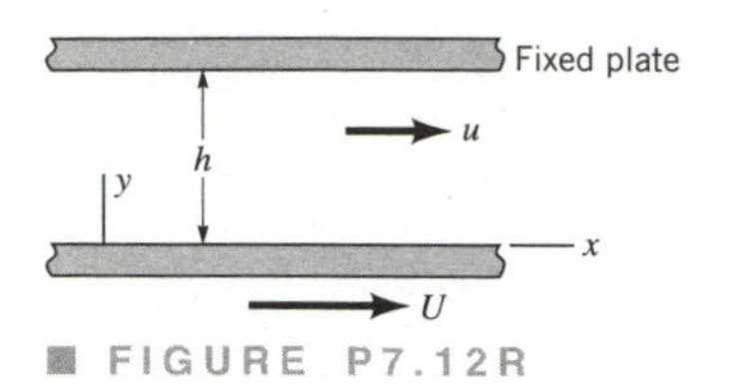

■ FIGURE P7.12R

Let $y^* = \frac{y}{h}$, $t^* = \frac{t}{\tau}$ (where $\tau = \frac{h^2\rho}{\mu}$), and $u^* = \frac{u}{U}$,
so that:

$$\frac{\partial u}{\partial t} = \frac{\partial(Uu^*)}{\partial t^*}\frac{\partial t^*}{\partial t} = U\frac{\partial u^*}{\partial t^*}\left(\frac{1}{\tau}\right) = \frac{U}{\tau}\frac{\partial u^*}{\partial t^*}$$

$$\frac{\partial u}{\partial y} = \frac{\partial(Uu^*)}{\partial y^*}\frac{\partial y^*}{\partial y} = U\frac{\partial u^*}{\partial y^*}\left(\frac{1}{h}\right) = \frac{U}{h}\frac{\partial u^*}{\partial y^*}$$

$$\frac{\partial^2 u}{\partial y^2} = \frac{U}{h}\frac{\partial}{\partial y^*}\left(\frac{\partial u^*}{\partial y^*}\right)\frac{\partial y^*}{\partial y} = \frac{U}{h^2}\frac{\partial^2 u^*}{\partial y^{*2}}$$

Thus, the original differential equation becomes

$$\left[\frac{\rho U}{\tau}\right]\frac{\partial u^*}{\partial t^*} = \left[\frac{\mu U}{h^2}\right]\frac{\partial^2 u^*}{\partial y^{*2}}$$

and with $\tau = h^2\rho/\mu$

$$\left[\frac{\mu U}{h^2}\right]\frac{\partial u^*}{\partial t^*} = \left[\frac{\mu U}{h^2}\right]\frac{\partial^2 u^*}{\partial y^{*2}}$$

so that

$$\frac{\partial u^*}{\partial t^*} = \frac{\partial^2 u^*}{\partial y^{*2}}$$

The initial condition $u=0$ at $t=0$ becomes $u^*=0$ at $t^*=0$, and the boundary conditions $u=U$ at $y=0$; $u=0$ at $y=h$, become $u^*=1.0$ at $y^*=0$ and $u^*=0$ at $y^*=1.0$.

7.13R

7.13R (Dimensionless governing equations) The flow between two concentric cylinders (see Fig. P7.13R) is governed by the differential equation

$$\frac{d^2 v_\theta}{dr^2} + \frac{d}{dr}\left(\frac{v_\theta}{r}\right) = 0$$

where v_θ is the tangential velocity at any radial location, r. The inner cylinder is fixed and the outer cylinder rotates with an angular velocity ω. Express the equation in dimensionless form using R_o and ω as reference parameters.

(ANS: $d^2v_\theta^*/dr^{*2} + d(v_\theta^*/r^*)dr^* = 0$)

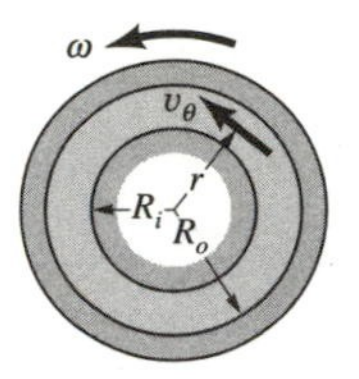

■ FIGURE P7.13R

Let $r^* = \frac{r}{R_o}$ and $v_\theta^* = \frac{v_\theta}{R_o \omega}$ so that

$$\frac{dv_\theta}{dr} = \frac{d(R_o \omega v_\theta^*)}{dr^*}\frac{dr^*}{dr} = R_o \omega \frac{dv_\theta^*}{dr^*}\left(\frac{1}{R_o}\right) = \omega \frac{dv_\theta^*}{dr^*}$$

and

$$\frac{d^2 v_\theta}{dr^2} = \omega \frac{d}{dr^*}\left(\frac{dv_\theta^*}{dr^*}\right)\frac{dr^*}{dr} = \frac{\omega}{R_o}\frac{d^2 v_\theta^*}{dr^{*2}}$$

Similarly,

$$\frac{d}{dr}\left(\frac{v_\theta}{r}\right) = \frac{d}{dr^*}\left(\frac{R_o \omega v_\theta^*}{R_o r^*}\right)\frac{dr^*}{dr} = \frac{\omega}{R_o}\frac{d}{dr^*}\left(\frac{v_\theta^*}{r^*}\right)$$

Thus, the original differential equation becomes

$$\left[\frac{\omega}{R_o}\right]\frac{d^2 v_\theta^*}{dr^{*2}} + \left[\frac{\omega}{R_o}\right]\frac{d}{dr^*}\left(\frac{v_\theta^*}{r^*}\right) = 0$$

or

$$\frac{d^2 v_\theta^*}{dr^{*2}} + \frac{d}{dr^*}\left(\frac{v_\theta^*}{r^*}\right) = 0$$

8

Viscous Flow in Pipes

Turbulent jet: The jet of water from the pipe is turbulent. The complex, irregular, unsteady structure typical of turbulent flows is apparent. (Laser-induced fluorescence of dye in water.) (Photography by P. E. Dimotakis, R. C. Lye, and D. Z. Papantoniou.)

8.1R

8.1R (Laminar flow) Asphalt at 120 °F, considered to be a Newtonian fluid with a viscosity 80,000 times that of water and a specific gravity of 1.09, flows through a pipe of diameter 2.0 in. If the pressure gradient is 1.6 psi/ft determine the flowrate assuming the pipe is **(a)** horizontal; **(b)** vertical with flow up.

(ANS: 4.69×10^{-3} ft^3/s; 3.30×10^{-3} ft^3/s)

If the flow is laminar, then $Q = \dfrac{\pi(\Delta p - \gamma \ell \sin\theta)D^4}{128\mu \ell}$ (1)

where $\gamma = SG\,\gamma_{H_2O} = 1.09(62.4\frac{lb}{ft^3}) = 68.0\frac{lb}{ft^3}$ and from Table B.1 for μ_{H_2O} at 120°F,

$$\mu = 80{,}000\,\mu_{H_2O}\Big|_{120°F} = 8\times10^4(1.164\times10^{-5}\ \frac{lb\cdot s}{ft^2}) = 0.931\ \frac{lb\cdot s}{ft^2}$$

a) For horizontal flow, $\theta = 0$

Q → 2 in.

Thus, from Eq.(1)

$$Q = \frac{\pi(1.6\times144\ \frac{lb}{ft^2})(\frac{2}{12}ft)^4}{128(0.931\ \frac{lb\cdot s}{ft^2})(1\,ft)} = \underline{\underline{4.69\times10^{-3}\ \frac{ft^3}{s}}}$$

b) For vertical flow up, $\theta = 90$

g, Q

Thus, from Eq.(1)

$$Q = \frac{\pi(1.6\times144\ \frac{lb}{ft^2} - 68\frac{lb}{ft^3}(1\,ft))(\frac{2}{12}ft)^4}{128(0.931\ \frac{lb\cdot s}{ft^2})(1\,ft)} = \underline{\underline{3.30\times10^{-3}\ \frac{ft^3}{s}}}$$

Note: We must check to see if our assumption of laminar flow is correct.

Since $V = \frac{Q}{A} = \dfrac{4.69\times10^{-3}\frac{ft^3}{s}}{\frac{\pi}{4}(\frac{2}{12})^2} = 0.215\ \frac{ft}{s}$ it follows that

$$Re = \frac{\rho VD}{\mu} = \frac{1.09(1.94\frac{slug}{ft^3})(0.215)(\frac{2}{12}ft)}{0.931\ \frac{lb\cdot s}{ft^2}} = 0.0814 < 2100$$

The flow is laminar when $\theta = 0$. It is laminar for $\theta = 90°$ also.

8.2R

8.2R (Laminar flow) A fluid flows through two horizontal pipes of equal length which are connected together to form a pipe of length 2ℓ. The flow is laminar and fully developed. The pressure drop for the first pipe is 1.44 times greater than it is for the second pipe. If the diameter of the first pipe is D, determine the diameter of the second pipe.

(ANS: 1.095 D)

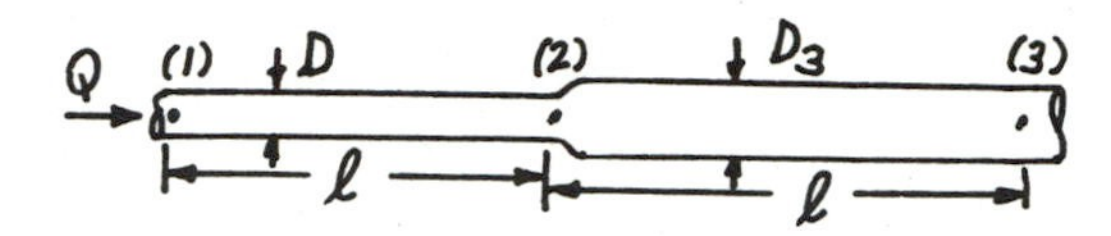

For laminar flow, $Q = \dfrac{\pi D^4 \Delta p}{128 \mu \ell}$, where $Q_1 = Q_3$ and $\Delta p_{1-2} = 1.44 \Delta p_{2-3}$

Thus,

$$Q_1 = \frac{\pi D^4 \Delta p_{1-2}}{128 \mu \ell} = Q_3 = \frac{\pi D_3^4 \Delta p_{2-3}}{128 \mu \ell}$$

or

$$D_3 = D\left(\frac{\Delta p_{1-2}}{\Delta p_{2-3}}\right)^{1/4} = D(1.44)^{1/4} = \underline{\underline{1.095\,D}}$$

8.3R

8.3R (Velocity profile) A fluid flows through a pipe of radius R with a Reynolds number of 100,000. At what location, r/R, does the fluid velocity equal the average velocity? Repeat if the Reynolds number is 1000.

(ANS: 0.758; 0.707)

For $Re = 10^5$, $\frac{\bar{u}}{V_c} = (1-\frac{r}{R})^{\frac{1}{n}}$, where $n = 7.1$ (see Fig. 8.17)
The relationship between the average velocity, V, and the centerline velocity, V_c, can be obtained from

$$Q = AV = \int \bar{u}\, dA \quad \text{or}$$

$$\pi R^2 V = \int \bar{u}\,(2\pi r\, dr) = 2\pi V_c \int_{r=0}^{r=R} (1-\frac{r}{R})^{\frac{1}{n}} r\, dr \qquad (1)$$

But, with

$x \equiv 1 - \frac{r}{R}$ so that $dx = -\frac{1}{R}dr$ and $r = R(1-x)$, we have

$$\int_{r=0}^{r=R} (1-\frac{r}{R})^{\frac{1}{n}} r\, dr = -R^2 \int_{x=1}^{x=0} x^{\frac{1}{n}}(1-x)\,dx = R^2 \int_0^1 [x^{\frac{1}{n}} - x^{\frac{n+1}{n}}]\,dx$$

$$= R^2 \left[\frac{n}{(n+1)} x^{\frac{n+1}{n}} - \frac{n}{(2n+1)} x^{\frac{2n+1}{n}}\right]_0^1 = R^2 \frac{n^2}{(n+1)(2n+1)} \qquad (2)$$

Thus, from Eqs. (1) and (2):

$$\pi R^2 V = 2\pi V_c R^2 \frac{n^2}{(n+1)(2n+1)} \quad \text{or} \quad V = \frac{2n^2}{(n+1)(2n+1)} V_c$$

Hence, $\bar{u} = V$ when

$$\bar{u} = V_c (1-\frac{r}{R})^{\frac{1}{n}} \quad \text{or} \quad V = \frac{(n+1)(2n+1)}{2n^2} V (1-\frac{r}{R})^{\frac{1}{n}}$$

or

$$\frac{r}{R} = 1 - \left[\frac{2n^2}{(n+1)(2n+1)}\right]^n = 1 - \left[\frac{2(7.1)^2}{(7.1+1)(2\times 7.1+1)}\right]^{7.1} = \underline{\underline{0.758}}$$

For $Re = 10^3$ the flow is laminar with $u = V_c[1-(\frac{r}{R})^2]$
and
$V = \frac{1}{2}V_c$

Thus, $u = V$ when

$$\frac{1}{2}V_c = V_c[1-(\frac{r}{R})^2], \text{ or } \frac{r}{R} = \sqrt{\frac{1}{2}} = \underline{\underline{0.707}}$$

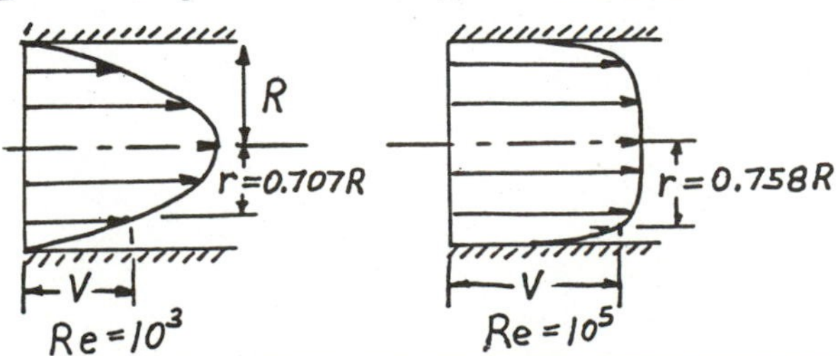

8.4R

8.4R (Turbulent velocity profile) Water at 80 °C flows through a 120-mm-diameter pipe with an average velocity of 2 m/s. If the pipe wall roughness is small enough so that it does not protrude through the laminar sublayer, the pipe can be considered as smooth. Approximately what is the largest roughness allowed to classify this pipe as smooth?

(ANS: 2.31×10^{-5} m)

Let h = roughness. Thus, $h = \delta_s$, where $\delta_s = \frac{5\nu}{u^*}$ with $u^* = \left(\frac{\tau_w}{\rho}\right)^{1/2}$ and $\tau_w = \frac{D\Delta p}{4\ell}$. Since $\Delta p = f\frac{\ell}{D}\frac{1}{2}\rho V^2$ we obtain

$$\tau_w = \frac{\rho f V^2}{8} \quad \text{or} \quad u^* = \left(\frac{fV^2}{8}\right)^{1/2}$$

With $Re = \frac{VD}{\nu} = \frac{(2\frac{m}{s})(0.12\,m)}{3.65\times10^{-7}\frac{m^2}{s}} = 6.58\times10^5$ and a smooth pipe we obtain $f = 0.0125$ (see Fig. 8.20).

Hence, $u^* = \left(\frac{0.0125(2\frac{m}{s})^2}{8}\right)^{1/2} = 0.0791\,\frac{m}{s}$

and

$$\delta_s = \frac{5(3.65\times10^{-7}\frac{m^2}{s})}{0.0791\frac{m}{s}} = \underline{\underline{2.31\times10^{-5}\,m}}$$

$R = D/2 = 60\,mm$ $\quad \nu\big|_{80°C} = 3.65\times10^{-7}\frac{m^2}{s}$ (see Table B.2)

h $\quad \delta_s = 0.0231\,mm$ = laminar sublayer thickness.

If the roughness element is smaller than 0.0231 mm it lies within the laminar sublayer.

8.5R

8.5R (Moody chart) Water flows in a smooth plastic pipe of 200-mm diameter at a rate of 0.10 m^3/s. Determine the friction factor for this flow.

(ANS: 0.0128)

For a $D = 0.200\ m$ plastic pipe, $\frac{\varepsilon}{D} = 0$ (see Table 8.1)

Also,

$$Re = \frac{VD}{\nu}, \text{ where } V = \frac{Q}{A} = \frac{0.10\ \frac{m^3}{s}}{\frac{\pi}{4}(0.2m)^2} = 3.18\ \frac{m}{s}$$

Hence,

$$Re = \frac{(3.18\frac{m}{s})(0.2m)}{1.12\times10^{-6}\frac{m^2}{s}} = 5.68\times10^5$$, so from Fig. 8.20 we obtain

$f = \underline{\underline{0.0128}}$

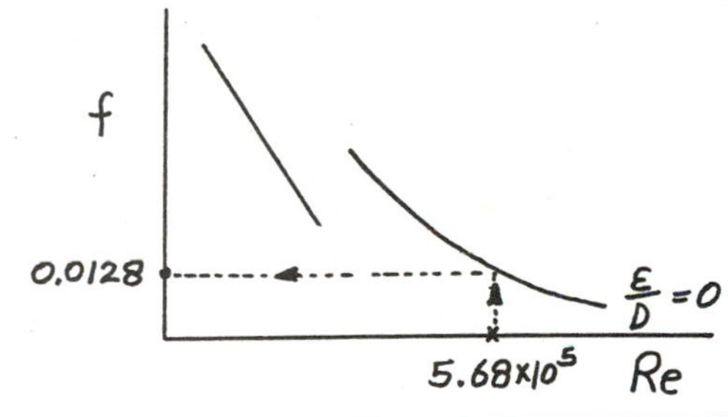

8.6R

8.6R (Moody chart) After a number of years of use, it is noted that to obtain a given flowrate, the head loss is increased to 1.6 times its value for the originally smooth pipe. If the Reynolds number is 10^6, determine the relative roughness of the old pipe.

(ANS: 0.00070)

Let $(\ )_o$ denote the old pipe and $(\ )_s$ the smooth pipe. Thus,

$\left(\frac{\varepsilon}{D}\right)_s = 0$, $Q_o = Q_s$, $V_o = V_s$ and $h_{L_o} = 1.6\ h_{L_s}$. Hence, with $D_o = D_s$

$f_o \frac{\ell_o}{D_o}\frac{V_o^2}{2g} = f_s \frac{\ell_s}{D_s}\frac{V_s^2}{2g}$, or $f_o = 1.6\ f_s$, where from Fig. 8.20 with $Re = 10^6$

$f_s = 0.0115$ Thus, $f_o = 1.6\ (0.0115) = 0.0184$, which with $Re = 10^6$ implies

$\left(\frac{\varepsilon}{D}\right)_o = \underline{\underline{0.00070}}$

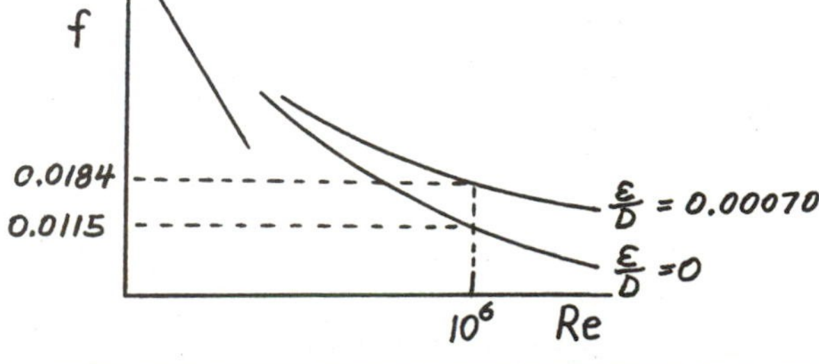

8.7R

8.7R (Minor losses) Air flows through the fine mesh gauze shown in Fig. P8.7R with an average velocity of 1.50 m/s in the pipe. Determine the loss coefficient for the gauze.
(ANS: 56.7)

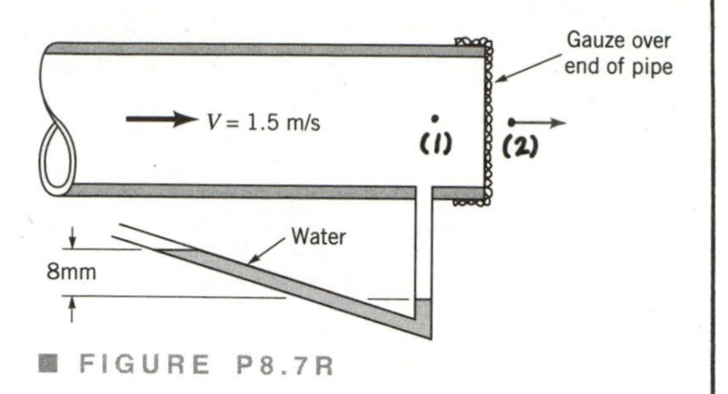

FIGURE P8.7R

$$\frac{p_1}{\gamma}+\frac{V_1^2}{2g}+z_1=\frac{p_2}{\gamma}+\frac{V_2^2}{2g}+z_2+K_L\frac{V^2}{2g}, \text{ where } z_1=z_2,\ V_1=V_2=V=1.5\,\frac{m}{s}$$

Thus, $K_L=\frac{2(p_1-p_2)}{\rho V^2}$

where $p_2=0$ and $p_1=8\text{mm water}$
or $p_1=(8\times10^{-3}\,m)(9.80\times10^3\,\frac{N}{m^3})=78.4\,\frac{N}{m^2}$

Hence,
$$K_L=\frac{2\,(78.4\,\frac{N}{m^2})}{(1.23\,\frac{kg}{m^3})(1.5\,\frac{m}{s})^2}=\underline{\underline{56.7}}$$

8.8R

8.8R (Noncircular conduits) A manufacturer makes two types of drinking straws: one with a square cross-sectional shape, and the other type the typical round shape. The amount of material in each straw is to be the same. That is, the length of the perimeter of the cross section of each shape is the same. For a given pressure drop, what is the ratio of the flowrates through the straws? Assume the drink is viscous enough to ensure laminar flow and neglect gravity.

(ANS: $Q_{round} = 1.83\, Q_{square}$)

(1) circle of diameter D (2) square of side a

$\Delta p_1 = \Delta p_2$ where $\Delta p = \gamma h_L$

Since $h_{L_1} = h_{L_2}$,

$$f_1 \frac{\ell_1}{D_{h_1}} \frac{V_1^2}{2g} = f_2 \frac{\ell_2}{D_{h_2}} \frac{V_2^2}{2g}$$, where $\ell_1 = \ell_2$, $D_{h_1} = D$, and $D_{h_2} = \frac{4A_2}{P_2} = \frac{4a^2}{4a} = a$

Thus,

$$\frac{f_1 V_1^2}{D} = \frac{f_2 V_2^2}{a} \qquad (1)$$

Since the perimeters are equal, $P_1 = P_2$, or $\pi D = 4a$. Hence, $a = \frac{\pi}{4} D$

For laminar flow $f = \frac{C}{Re_h}$, where $C_1 = 64$ and $C_2 = 56.9$ (Table 8.3)

and $Re_{h_1} = \frac{V_1 D_{h1}}{\nu} = \frac{V_1 D}{\nu}$, $Re_{h_2} = \frac{V_2 D_{h2}}{\nu} = \frac{V_2 a}{\nu} = \frac{\pi V_2 D}{4\nu}$

Thus, from Eq.(1)

$$\frac{\frac{64}{\left(\frac{V_1 D}{\nu}\right)} V_1^2}{D} = \frac{\frac{56.9}{\left(\frac{\pi V_2 D}{4\nu}\right)} V_2^2}{\frac{\pi D}{4}} \quad \text{or} \quad V_1 = 1.441\, V_2$$

Also, $Q_1 = A_1 V_1 = \frac{\pi}{4} D^2 V_1$ and $Q_2 = A_2 V_2 = a^2 V_2 = \frac{\pi^2}{16} D^2 V_2$

so that

$$\frac{Q_1}{Q_2} = \frac{\frac{\pi}{4} D^2 V_1}{\frac{\pi^2}{16} D^2 V_2} = \frac{4}{\pi} \frac{V_1}{V_2} = \frac{4}{\pi}(1.441) = 1.83$$

$$Q_{round} = 1.83\, Q_{square}$$

8.9R (Single pipe—determine pressure drop) Determine the pressure drop per 300-m length of new 0.20-m-diameter horizontal cast iron water pipe when the average velocity is 1.7 m/s.

(ANS: 47.6 kN/m²)

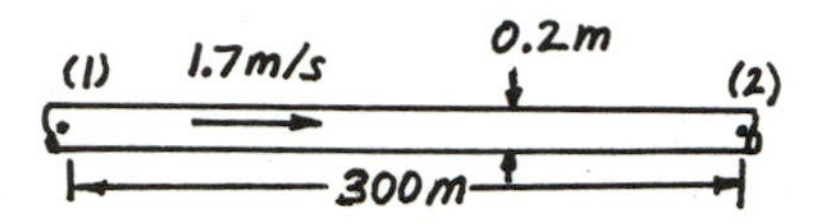

$$\frac{p_1}{\gamma} + \frac{V_1^2}{2g} + z_1 = \frac{p_2}{\gamma} + \frac{V_2^2}{2g} + z_2 + f\frac{\ell}{D}\frac{V^2}{2g}$$

or with $V_1 = V_2 = V = 1.7\,\frac{m}{s}$, $\ell = 300m$, $D = 0.2m$ this gives

$\Delta p = p_1 - p_2 = \gamma(z_2 - z_1) + f\frac{\ell}{D}\frac{1}{2}\rho V^2 = f\frac{\ell}{D}\frac{1}{2}\rho V^2$ if the pipe is horizontal.

But from Table 8.1, $\frac{\varepsilon}{D} = \frac{0.26\times10^{-3}\,m}{0.2\,m} = 1.3\times10^{-3}$

Also, $Re = \frac{VD}{\nu} = \frac{(1.7\,\frac{m}{s})(0.2\,m)}{1.12\times10^{-6}\,\frac{m^2}{s}} = 3.04\times10^5$

so that from Fig. 8.20, $f = 0.022$

Thus,

$$\Delta p = 0.022\left(\frac{300\,m}{0.2\,m}\right)\frac{1}{2}\left(999\,\frac{kg}{m^3}\right)\left(1.7\,\frac{m}{s}\right)^2 = 4.76\times10^4\,\frac{N}{m^2} = \underline{\underline{47.6\,\frac{kN}{m^2}}}$$

8.10R (Single pipe—determine pressure drop) A fire protection association code requires a minimum pressure of 65 psi at the outlet end of a 250-ft-long, 4-in.-diameter hose when the flowrate is 500 gal/min. What is the minimum pressure allowed at the pumper truck that supplies water to the hose? Assume a roughness of $\varepsilon = 0.03$ in.

(ANS: 94.0 psi)

(1) $D = 4$ in. $\varepsilon = 0.03$ in. (2)

$Q = 500$ gal/min $\ell = 250$ ft $p_2 = 65$ psi

$$\frac{p_1}{\gamma} + \frac{V_1^2}{2g} + z_1 = \frac{p_2}{\gamma} + \frac{V_2^2}{2g} + z_2 + f\frac{\ell}{D}\frac{V^2}{2g}, \text{ where } z_1 = z_2,\ V_1 = V_2 = V$$

and $Q = 500 \frac{gal}{min}\left(231 \frac{in.^3}{gal}\right)\left(\frac{1\ ft^3}{1728\ in.^3}\right)\left(\frac{1\ min}{60\ s}\right) = 1.114 \frac{ft^3}{s}$

Thus,

$$p_1 = p_2 + f\frac{\ell}{D}\frac{1}{2}\rho V^2, \text{ where } V = \frac{Q}{A} = \frac{1.114 \frac{ft^3}{s}}{\frac{\pi}{4}\left(\frac{4}{12} ft\right)^2} = 12.8 \frac{ft}{s} \qquad (1)$$

From Fig. 8.20 with

$$Re = \frac{VD}{\nu} = \frac{\left(12.8 \frac{ft}{s}\right)\left(\frac{4}{12} ft\right)}{1.21 \times 10^{-5} \frac{ft^2}{s}} = 3.53 \times 10^5 \text{ and } \frac{\varepsilon}{D} = \frac{0.03\ in.}{4\ in.} = 0.0075$$

we obtain

$f = 0.035$ Hence, from Eq. (1)

$$p_1 = 65 \frac{lb}{in.^2} + (0.035)\left(\frac{250\ ft}{\frac{4}{12}\ ft}\right)\frac{\left(12.8 \frac{ft}{s}\right)^2}{2}\left(1.94 \frac{slugs}{ft^3}\right)\left(\frac{1\ ft^2}{144\ in.^2}\right)$$

or

$$p_1 = 65\ psi + 29.0\ psi = \underline{\underline{94.0\ psi}}$$

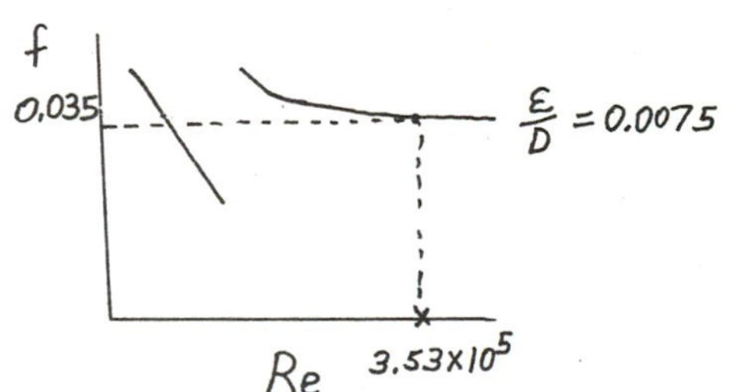

8.11R

8.11R (Single pipe—determine flowrate) An above ground swimming pool of 30 ft diameter and 5 ft depth is to be filled from a garden hose (smooth interior) of length 100 ft and diameter 5/8 in. If the pressure at the faucet to which the hose is attached remains at 55 psi, how long will it take to fill the pool? The water exits the hose as a free jet 6 ft above the faucet.

(ANS: 32.0 hr)

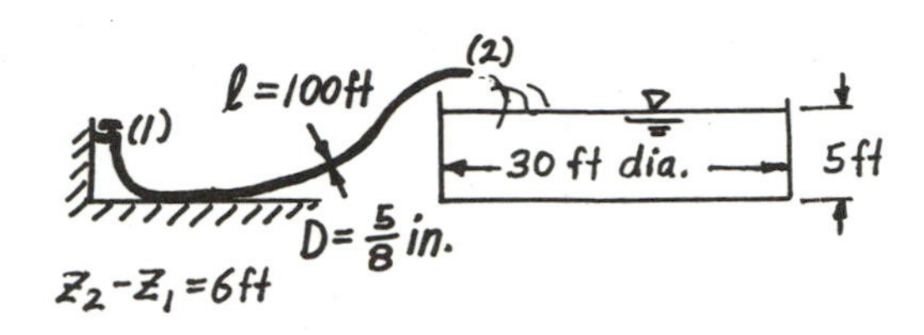

$\frac{p_1}{\gamma} + \frac{V_1^2}{2g} + z_1 = \frac{p_2}{\gamma} + \frac{V_2^2}{2g} + z_2 + f\frac{\ell}{D}\frac{V^2}{2g}$, where $p_1 = 55\frac{lb}{in.^2}$, $p_2 = 0$, $z_1 = 0$,

$z_2 = 6\,ft$, and $V_1 = V_2 = V$

Thus,

$$\frac{p_1}{\gamma} = z_2 + f\frac{\ell}{D}\frac{V^2}{2g} \quad \text{or} \quad \frac{(55\frac{lb}{in.^2})(144\frac{in.^2}{ft^2})}{62.4\frac{lb}{ft^3}} = 6\,ft + f\left(\frac{100\,ft}{\frac{5}{8(12)}\,ft}\right)\frac{V^2}{2(32.2\frac{ft}{s^2})}$$

or

$$fV^2 = 4.06, \text{ where } V \sim ft \tag{1}$$

Also, $Re = \frac{VD}{\nu} = \frac{(\frac{5}{8(12)}\,ft)V}{1.21\times10^{-5}\frac{ft^2}{s}}$, or $Re = 4300V$, where $V \sim ft$ (2)

and from Fig. 8.20 with $\frac{\varepsilon}{D} = 0$

f

Re

(3)

Trial and error solution of Eqs. (1), (2), and (3) for f, V, and Re:

Assume $f = 0.02$; from Eq. (1) $V = 14.2\frac{ft}{s}$; from Eq. (2) $Re = 6.11\times10^4$, so from Eq. (3) $f = 0.0196 \neq 0.02$ (but close!)

Assume $f = 0.0196$; from Eq. (1) $V = 14.4\frac{ft}{s}$; from Eq. (2) $Re = 6.19\times10^4$, so from Eq. (3) $f = 0.0196$, which agrees with the assumed value.

Thus, $V = 14.4\frac{ft}{s}$, or $Q = AV = \frac{\pi}{4}\left(\frac{5}{8(12)}\,ft\right)^2\left(14.4\frac{ft}{s}\right) = 0.0307\frac{ft^3}{s}$

Also, $\mathcal{V} = Qt$, where t = filling time and $\mathcal{V}$ = pool volume,

or

$$t = \frac{\mathcal{V}}{Q} = \frac{\frac{\pi}{4}(30\,ft)^2(5\,ft)}{0.0307\frac{ft^3}{s}} = 1.151\times10^5\,s\left(\frac{1\,hr}{3600\,s}\right) = \underline{\underline{32.0\,hr}}$$

8.12R

8.12R (Single pipe—determine pipe diameter) Water is to flow at a rate of 1.0 m^3/s through a rough concrete pipe ($\varepsilon = 3$ mm) that connects two ponds. Determine the pipe diameter if the elevation difference between the two ponds is 10 m and the pipe length is 1000 m. Neglect minor losses.

(ANS: 0.748 m)

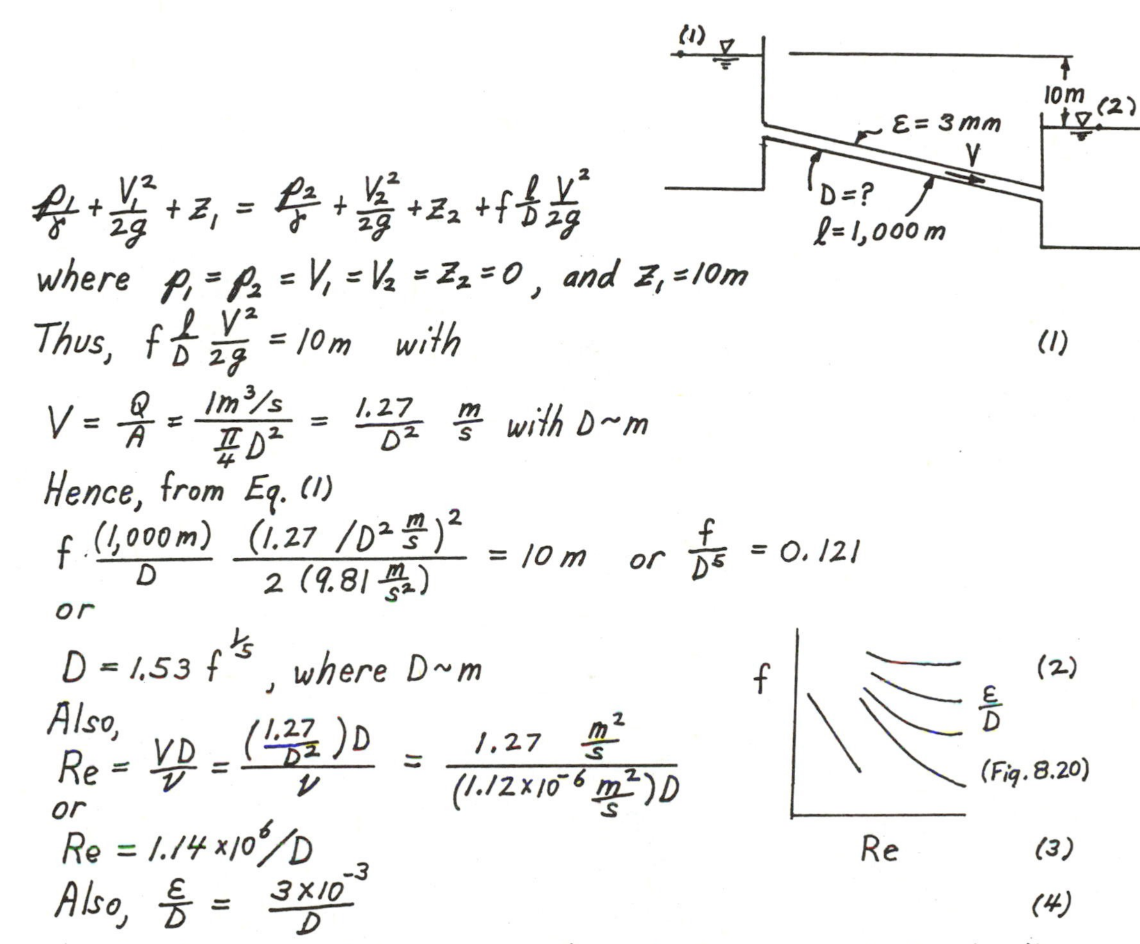

$$\frac{p_1}{\gamma} + \frac{V_1^2}{2g} + z_1 = \frac{p_2}{\gamma} + \frac{V_2^2}{2g} + z_2 + f\frac{\ell}{D}\frac{V^2}{2g}$$

where $p_1 = p_2 = V_1 = V_2 = z_2 = 0$, and $z_1 = 10m$

Thus, $f\frac{\ell}{D}\frac{V^2}{2g} = 10m$ with (1)

$$V = \frac{Q}{A} = \frac{1m^3/s}{\frac{\pi}{4}D^2} = \frac{1.27}{D^2}\ \frac{m}{s} \text{ with } D \sim m$$

Hence, from Eq. (1)

$$f \cdot \frac{(1{,}000m)}{D}\ \frac{(1.27\ /D^2 \frac{m}{s})^2}{2\ (9.81 \frac{m}{s^2})} = 10\ m \quad \text{or} \quad \frac{f}{D^5} = 0.121$$

or

$$D = 1.53\ f^{1/5}, \text{ where } D \sim m \qquad (2)$$

Also,

$$Re = \frac{VD}{\nu} = \frac{(\frac{1.27}{D^2})D}{\nu} = \frac{1.27\ \frac{m^2}{s}}{(1.12 \times 10^{-6}\ \frac{m^2}{s})D}$$

or

$$Re = 1.14 \times 10^6 / D \qquad (3)$$

Also, $\frac{\varepsilon}{D} = \frac{3 \times 10^{-3}}{D}$ (4)

Trial and error solution: 4 equations ((2), (3), (4), and Moody chart) and 4 unknowns (Re, D, f, and ε/D)

Assume $f = 0.02$ so that from Eq. (1), $D = 1.53\,(0.02)^{1/5} = 0.700m$; from Eq. (2), $Re = 1.63 \times 10^6$ and from Eq. (4), $\frac{\varepsilon}{D} = \frac{3 \times 10^{-3}}{0.700} = 4.29 \times 10^{-3}$

Thus, from the Moody chart (Fig. 8.20),

$f = 0.029 \neq 0.02$, the assumed value. Try again.

Assume $f = 0.029$ which gives $D = 0.754m$, $Re = 1.51 \times 10^6$, and $\frac{\varepsilon}{D} = 3.98 \times 10^{-3}$. Hence, from Fig. 8.20, $f = 0.028 \neq 0.029$.

Another try with $f = 0.028$ will show that this is the correct value.

Thus, $D = 1.53\,(0.028)^{1/5} = \underline{\underline{0.748\ m}}$

8.13R

8.13R (Single pipe with pump) Without the pump shown in Fig. P8.13R it is determined that the flowrate is too small. Determine the horsepower added to the fluid if the pump causes the flowrate to be doubled. Assume the friction factor remains at 0.020 in either case.

(ANS: 1.51 hp)

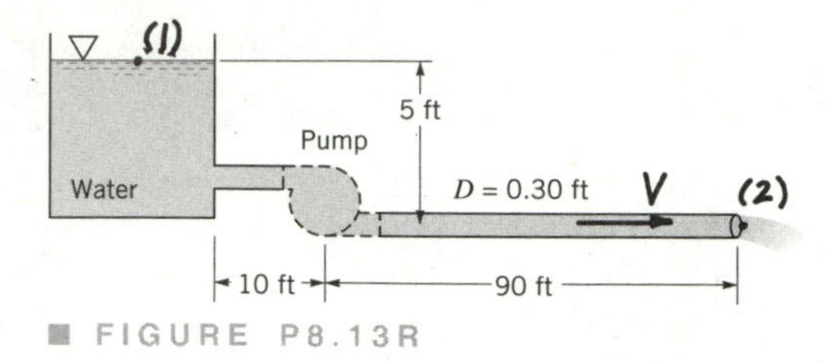

■ FIGURE P8.13R

$$h_p + \frac{p_1}{\gamma} + \frac{V_1^2}{2g} + z_1 = \frac{p_2}{\gamma} + \frac{V_2^2}{2g} + z_2 + \left(f\frac{\ell}{D} + K_L\right)\frac{V^2}{2g}$$, where $p_1 = p_2 = 0$, $V_1 = 0$, $z_1 = 5\,ft$,

$z_2 = 0$, and $V_2 = V$. Assume a sharp edged entrance: $K_L = 0.5$ (Fig. 8.22)

Without the pump, $h_p = 0$ Thus,

$$z_1 = \left(1 + f\frac{\ell}{D} + 0.5\right)\frac{V^2}{2g} \quad \text{or} \quad 5\,ft = \left(1 + 0.02\left(\frac{100\,ft}{0.3\,ft}\right) + 0.5\right)\frac{V^2}{2(32.2\,\frac{ft}{s^2})}$$

Thus, $V = 6.28\,\frac{ft}{s}$

With the pump $V = 2(6.28\,\frac{ft}{s}) = 12.56\,\frac{ft}{s}$

Thus, $h_p + z_1 = \left(1 + f\frac{\ell}{D} + 0.5\right)\frac{V^2}{2g}$

or

$$h_p = -5\,ft + \left(1 + 0.02\left(\frac{100\,ft}{0.3\,ft}\right) + 0.5\right)\frac{(12.56\,\frac{ft}{s})^2}{2(32.2\,\frac{ft}{s^2})} = -5\,ft + 20\,ft = 15\,ft$$

Hence, from Eq. 5.85

$$\dot{W}_p = \gamma Q h_p = (62.4\,\frac{lb}{ft^3})(\frac{\pi}{4})(0.3\,ft)^2(12.56\,\frac{ft}{s})(15\,ft) = 831\,\frac{ft\cdot lb}{s}\left(\frac{1\,hp}{550\,\frac{ft\cdot lb}{s}}\right)$$

or

$\dot{W}_p = \underline{\underline{1.51\,hp}}$

8.14R

8.14R (Single pipe with pump) The pump shown in Fig. P8.14R adds a 15-ft head to the water being pumped when the flowrate is 1.5 ft^3/s. Determine the friction factor for the pipe.
(ANS: 0.0306)

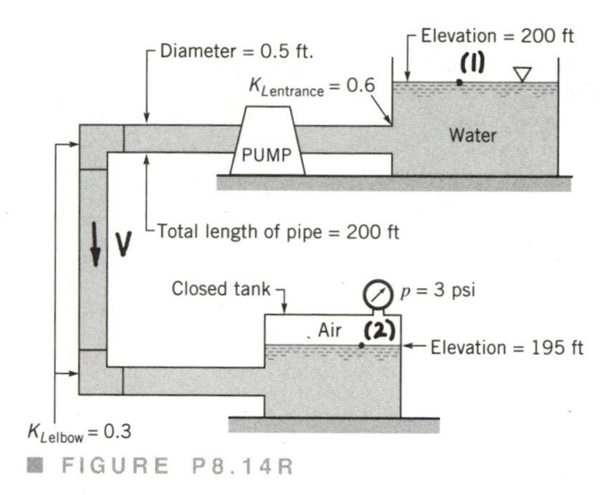

FIGURE P8.14R

For flow from the upper tank to the lower tank:

$$\frac{p_1}{\gamma}+\frac{V_1^2}{2g}+z_1+h_p=\frac{p_2}{\gamma}+\frac{V_2^2}{2g}+z_2+\left(f\frac{\ell}{D}+\Sigma K_L\right)\frac{V^2}{2g} \qquad (1)$$

where $p_1=0$, $V_1=0$, $z_1=200\text{ ft}$, $h_p=15\text{ ft}$, $z_2=195\text{ ft}$, $V_2=0\frac{ft}{s}$, and

$\Sigma K_L = K_{L_{ent}}+2K_{L_{elbow}}+K_{L_{exit}} = 0.6+2(0.3)+1=2.2$ (see Fig. 8.25)

Thus, Eq. (1) becomes

$$200\text{ ft}+15\text{ ft}=\frac{(3\frac{lb}{in.^2})(144\frac{in.^2}{ft^2})}{62.4\frac{lb}{ft^3}}+195\text{ ft}+\left(f\frac{(200\text{ ft})}{(0.5\text{ ft})}+2.2\right)\frac{V^2}{2(32.2\frac{ft}{s^2})} \qquad (2)$$

but with

$V=\frac{Q}{A}=\frac{1.5\frac{ft^3}{s}}{\frac{\pi}{4}(0.5\text{ ft})^2}=7.64\frac{ft}{s}$, Eq. (2) gives $f=\underline{\underline{0.0306}}$

Note: If the flow was from the lower tank (2) to the upper tank (1), then

$$\frac{p_2}{\gamma}+\frac{V_2^2}{2g}+z_2+h_p=\frac{p_1}{\gamma}+\frac{V_1^2}{2g}+z_1+\left[f\frac{\ell}{D}+\Sigma K_L\right]\frac{V^2}{2g}$$

or

$$\frac{(3\frac{lb}{in.^2})(144\frac{in.^2}{ft^2})}{62.4\,lb/ft^3}+195\text{ ft}+15\text{ ft}=200\text{ ft}+\left[f\frac{(200\text{ ft})}{(0.5\text{ ft})}+2.2\right]\frac{(7.64\frac{ft}{s})^2}{2(32.2\,ft/s^2)}$$

which gives $f=\underline{\underline{0.0412}}$

Both the assumed downflow ($f=0.0306$) and the upflow ($f=0.0412$) give reasonable friction factor values.

8.15R

8.15R (Single pipe with turbine) Water drains from a pressurized tank through a pipe system as shown in Fig. P8.15R. The head of the turbine is equal to 116 m. If entrance effects are negligible, determine the flow rate.

(ANS: 3.71×10^{-2} m³/s)

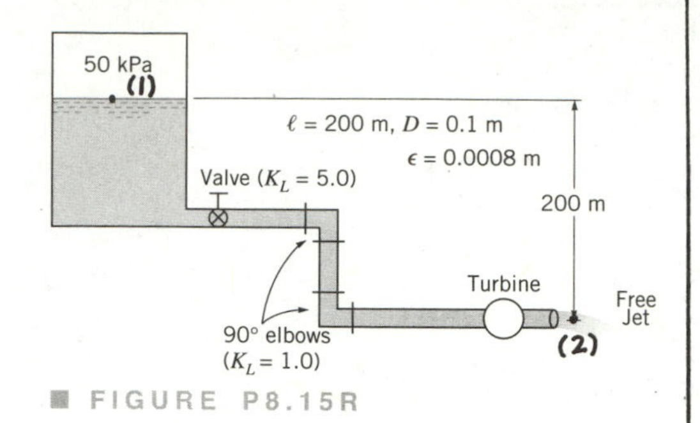

FIGURE P8.15R

$$\frac{p_1}{\gamma} + \frac{V_1^2}{2g} + z_1 + h_s = \frac{p_2}{\gamma} + \frac{V_2^2}{2g} + z_2 + \left[f\frac{\ell}{D} + \Sigma K_L\right]\frac{V^2}{2g}$$

where $z_1 = 200\,m$,

$p_2 = V_1 = z_2 = 0$, and $h_s = -116\,m$ ($h_s < 0$ since it is a turbine)

Thus,

$$\frac{50\times10^3\,\frac{N}{m^2}}{9.8\times10^3\,\frac{N}{m^3}} + 200\,m - 116\,m = \frac{V^2\,\frac{m^2}{s^2}}{2(9.81\frac{m}{s^2})}\left[1 + f\left(\frac{200m}{0.1m}\right) + 5 + 2(1)\right]$$

or

$$V^2 = \frac{1748}{(8+2000f)}, \text{ where } V \sim \frac{m}{s} \qquad (1)$$

Also,

$$Re = \frac{\rho V D}{\mu} = \frac{(999\frac{kg}{m^3})V\frac{m}{s}(0.1m)}{1.12\times10^{-3}\frac{N\cdot s}{m^2}}, \text{ or } Re = 8.92\times10^4\,V \qquad (2)$$

The final equation is the Moody chart (Eq. 8.20) with

$$\frac{\varepsilon}{D} = \frac{0.0008\,m}{0.1\,m} = 0.008$$

f

$\frac{\varepsilon}{D} = 0.008$ (3)

Re

Trial and error solution:
(3 equations; 3 unknowns: Re, f, V)

Assume $f = 0.038$ so that from Eq. (1), $V = 4.56\,\frac{m}{s}$; from Eq. (2), $Re = 4.07\times10^5$; and from the Moody chart, $f = 0.035$ which does not equal the assumed value. Try again.
Assume $f = 0.035$ which gives $V = 4.73\,\frac{m}{s}$ and $Re = 4.22\times10^5$. Hence, from the Moody chart $f = 0.035$ which agrees with the assumed value. Thus, $V = 4.73\,\frac{m}{s}$ and

$$Q = AV = \frac{\pi}{4}(0.1m)^2(4.73\frac{m}{s}) = \underline{\underline{3.71\times10^{-2}\,\frac{m^3}{s}}}$$

8.16R

8.16R (Multiple pipes) The three tanks shown in Fig. P8.16R are connected by pipes with friction factors of 0.03 for each pipe. Determine the water velocity in each pipe. Neglect minor losses.

(ANS: (A) 4.73 ft/s, (B) 8.35 ft/s, (C) 10.3 ft/s)

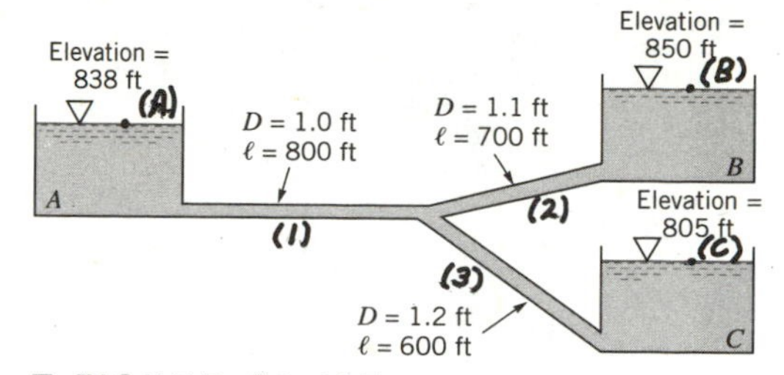

■ FIGURE P8.16R

Assume the flow from both tanks A and B is into tank C, or $Q_3 = Q_1 + Q_2$

Thus, $\frac{\pi}{4} D_3^2 V_3 = \frac{\pi}{4} D_1^2 V_1 + \frac{\pi}{4} D_2^2 V_2$, or $1.2^2 V_3 = 1.0^2 V_1 + 1.1^2 V_2$

Hence, $V_3 = 0.694\, V_1 + 0.840\, V_2$ (1)

For the flow from A to C, with $p_A = p_C = 0$, $V_A = V_C = 0$, we obtain

$z_A = z_C + f_1 \frac{\ell_1}{D_1} \frac{V_1^2}{2g} + f_3 \frac{\ell_3}{D_3} \frac{V_3^2}{2g}$, or $838\,\text{ft} = 805\,\text{ft} + \frac{0.03}{2(32.2\,\frac{\text{ft}}{\text{s}^2})}\left[\frac{800\,\text{ft}}{1\,\text{ft}} V_1^2 + \frac{600\,\text{ft}}{1.2\,\text{ft}} V_3^2\right]$

or

$33 = 0.373\, V_1^2 + 0.233\, V_3^2$ (2)

Similarly for the flow from B to C, with $p_B = p_C = 0$, $V_B = V_C = 0$, we obtain

$z_B = z_C + f_2 \frac{\ell_2}{D_2} \frac{V_2^2}{2g} + f_3 \frac{\ell_3}{D_3} \frac{V_3^2}{2g}$, or $850\,\text{ft} = 805\,\text{ft} + \frac{0.03}{2(32.2\,\frac{\text{ft}}{\text{s}^2})}\left[\frac{700\,\text{ft}}{1.1\,\text{ft}} V_2^2 + \frac{600\,\text{ft}}{1.2\,\text{ft}} V_3^2\right]$

or

$45 = 0.296\, V_2^2 + 0.233\, V_3^2$ (3)

Thus, 3 equations ((1), (2), and (3)) for V_1, V_2, and V_3. Solve as follows:

Subtract (2) from (3) to obtain

$12 = 0.296\, V_2^2 - 0.373\, V_1^2$ (4)

From (2): $V_3 = \sqrt{141.6 - 1.6\, V_1^2}$, or when combined with (1):

$\sqrt{141.6 - 1.6 V_1^2} = 0.694\, V_1 + 0.840\, V_2$, or $V_2 = \sqrt{200 - 2.27 V_1^2} - 0.826\, V_1$ (5)

Combine Eqs. (4) and (5) to obtain:

$\frac{12}{0.296} = \left[\sqrt{200 - 2.27 V_1^2} - 0.826\, V_1\right]^2 - \frac{0.373}{0.296}$, which can be simplified to

$V_1 \sqrt{200 - 2.27 V_1^2} = 96.5 - 1.725\, V_1^2$ By squaring this equation we (6)

obtain (after simplification):

$V_1^4 - 101.5\, V_1^2 + 1774 = 0$ Hence: $V_1^2 = \frac{101.5 \pm \sqrt{101.5^2 - 4(1774)}}{2} = 79.1$ or 22.4

Thus, $V_1 = 8.89\,\frac{\text{ft}}{\text{s}}$ or $V_1 = 4.73\,\frac{\text{ft}}{\text{s}}$

Note: The $V_1 = 8.89$ solution is an extra root introduced by squaring Eq. (6). It is not a solution of the original Eqs. (1), (2), (3). For this value, Eq. (6) becomes $8.89\sqrt{200 - 2.27(8.89^2)} \stackrel{?}{=} 96.5 - 1.725\,(8.89)^2$, or "$40 = -40$"

Thus $V_1 = \underline{\underline{4.73\,\frac{\text{ft}}{\text{s}}}}$, from Eq. (2) $V_3 = \left[\frac{33 - 0.373(4.73)^2}{0.233}\right]^{1/2} = \underline{\underline{10.3\,\frac{\text{ft}}{\text{s}}}}$,

and from Eq. (1) $V_2 = \frac{10.3 - 0.694(4.73)}{0.840} = \underline{\underline{8.35\,\frac{\text{ft}}{\text{s}}}}$

8.17R

8.17R (Flow meters) Water flows in a 0.10-m-diameter pipe at a rate of 0.02 m^3/s. If the pressure difference across the orifice meter in the pipe is to be 28 kPa, what diameter orifice is needed?

(ANS: 0.070 m)

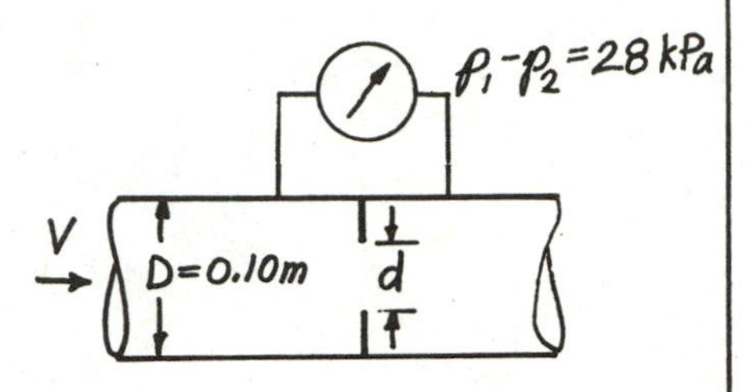

$$Q = C_o A_o \sqrt{\frac{2(p_1 - p_2)}{\rho(1-\beta^4)}}, \text{ where } Q = 0.02 \frac{m^3}{s},\ p_1 - p_2 = 28 \text{ kPa},\ \beta = \frac{d}{D},$$
and $\rho = 999 \frac{kg}{m^3}$.

Thus,

$$0.02 \frac{m^3}{s} = C_o \frac{\pi}{4} d^2 \sqrt{\frac{2(28 \times 10^3 \frac{N}{m^2})}{999 \frac{kg}{m^3}\left(1-\left(\frac{d}{0.1}\right)^4\right)}}, \text{ with } d \sim m$$

Hence,

$$3.40 \times 10^{-3} = \frac{C_o d^2}{\sqrt{1-(10d)^4}}, \text{ where } C_o = C_o(Re, \tfrac{d}{D}) \text{ from Fig. 8.41.} \qquad (1)$$

Also, $Re = \frac{VD}{\nu}$, where $V = \frac{Q}{\frac{\pi}{4}D^2} = \frac{0.02 \frac{m^3}{s}}{\frac{\pi}{4}(0.1m)^2} = 2.55 \frac{m}{s}$

Thus, $Re = \frac{(2.25 \frac{m}{s})(0.1m)}{1.12 \times 10^{-6} \frac{m^2}{s}} = 2.28 \times 10^5$

Trial and error solution: Assume a value of d; calculate $\beta = \frac{d}{D}$; look up C_o in Fig. 8.41 and calculate $\frac{C_o d^2}{\sqrt{1-(10d)^4}}$; compare with 3.40×10^{-3} (see Eq. (1)).

d, m	β	C_o	$\frac{C_o d^2}{\sqrt{1-(10d)^4}}$
0.070	0.70	0.61	3.43×10^{-3} ← checks
0.069	0.69	0.61	3.30×10^{-3}
0.071	0.71	0.61	3.56×10^{-3}

Thus, $d = \underline{\underline{0.070\,m}}$

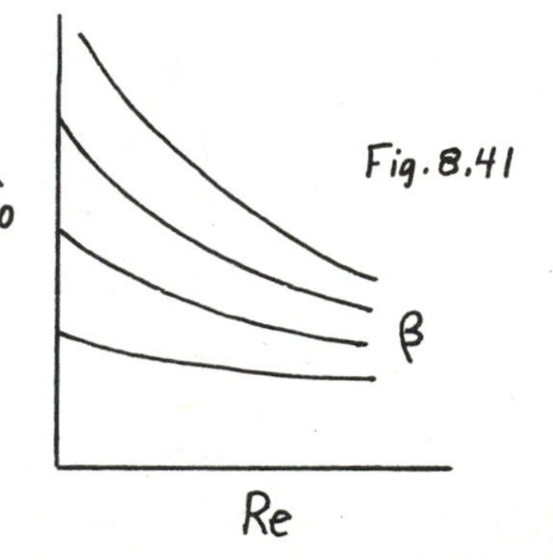

8.18R

8.18R (Flow meters) A 2.5-in.-diameter flow nozzle is installed in a 3.8-in.-diameter pipe that carries water at 160 °F. If the flowrate is 0.78 cfs, determine the reading on the inverted air-water U-tube manometer used to measure the pressure difference across the meter.

(ANS: 6.75 ft)

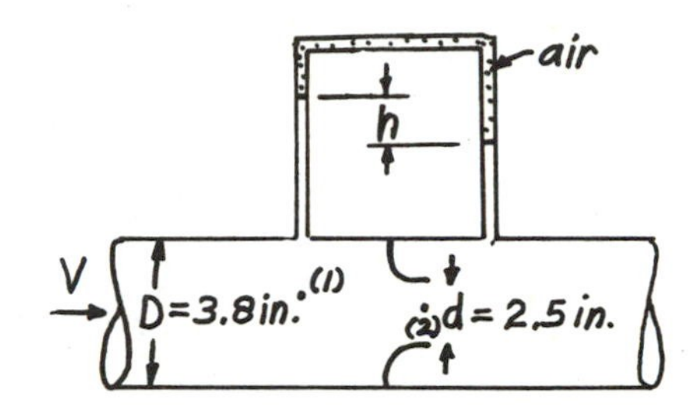

$$Q = C_n A_n \sqrt{\frac{2(p_1 - p_2)}{\rho(1-\beta^4)}}\text{, where } \beta = \frac{d}{D} = \frac{2.5\text{ in.}}{3.8\text{ in.}} = 0.658 \qquad (1)$$

$$V = \frac{Q}{\frac{\pi}{4}D^2} = \frac{0.78\frac{ft^3}{s}}{\frac{\pi}{4}\left(\frac{3.8}{12}ft\right)^2} = 9.90\frac{ft}{s}$$ and from Table B.1: $\rho = 1.896\frac{slugs}{ft^3}$

$\mu = 8.32\times10^{-6}\frac{lb\cdot s}{ft^2}$ at $T = 160°F$

Thus, with

$$Re = \frac{\rho V D}{\mu} = \frac{(1.896\frac{slugs}{ft^3})(9.90\frac{ft}{s})(\frac{3.8}{12}ft)}{8.32\times10^{-6}\frac{lb\cdot s}{ft^2}} = 7.14\times10^5$$ we obtain from Fig. 8.43:

$C_n = 0.989$ Thus, from Eq. (1)

$$0.78\frac{ft^3}{s} = (0.989)\frac{\pi}{4}\left(\frac{2.5}{12}ft\right)^2\sqrt{\frac{2(p_1-p_2)}{(1.896\frac{slugs}{ft^3})(1-0.658^4)}}$$

or

$$p_1 - p_2 = 412\frac{lb}{ft^2}$$

However, $p_1 - p_2 = \gamma_{H_2O}h$ so that $h = \frac{412\frac{lb}{ft^2}}{(32.2\frac{ft}{s^2})(1.896\frac{slugs}{ft^3})} = \underline{\underline{6.75\text{ ft}}}$

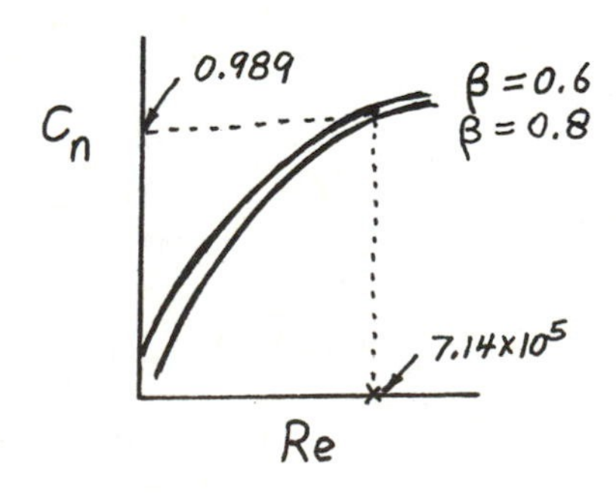

9

Flow Over Immersed Bodies

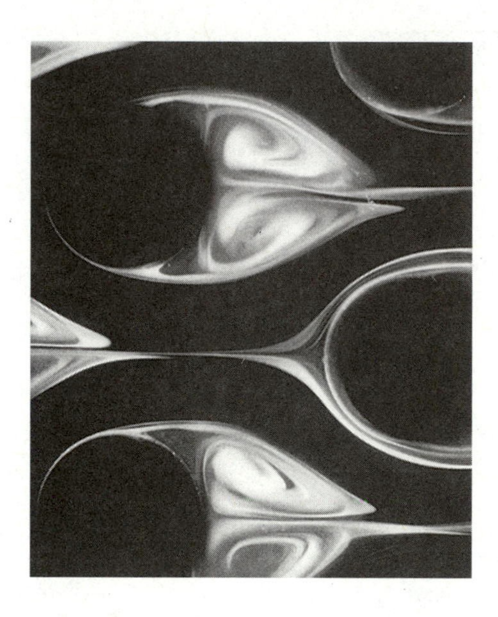

Impulsive start of flow past an array of cylinders: The complex structure of laminar flow past a relatively simple geometric structure illustrates why it is often difficult to obtain exact analytical results for external flows. (Dye in water.) (Photograph courtesy of ONERA, France.)

9.1R

9.1R (Lift/drag calculation) Determine the lift and drag coefficients (based on frontal area) for the triangular two-dimensional object shown in Fig. P9.1R. Neglect shear forces.
(ANS: 0; 1.70)

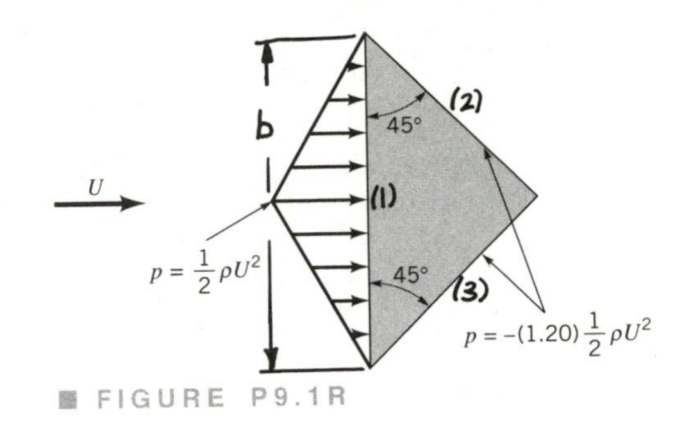

■ FIGURE P9.1R

$$\mathcal{D} = \int p \cos\theta \, dA + \int \tau_w \sin\theta \, dA, \text{ where } \tau_w = 0$$

Thus,

$$\mathcal{D} = \int_1 p \cos\theta \, dA + \int_2 p \cos\theta \, dA + \int_3 p \cos\theta \, dA$$

$$= \int_1 p \, dA - 2\int_2 p \cos 45° dA = \tfrac{1}{2}\left(\tfrac{1}{2}\rho U^2\right) \ell b - 2\left(-1.20\left(\tfrac{1}{2}\rho U^2\right)\cos 45° A_2\right)$$

where

$$A_2 = \ell \frac{b/2}{\cos 45°}$$

ℓ = length of object

Hence,

$$\mathcal{D} = 1.7\left(\tfrac{1}{2}\rho U^2\right)\ell b$$

or

$$C_D = \frac{\mathcal{D}}{\frac{1}{2}\rho U^2 A} = \frac{1.7\left(\frac{1}{2}\rho U^2\right)\ell b}{\frac{1}{2}\rho U^2 \ell b} = \underline{\underline{1.70}}$$

Because of symmetry of the object, $\mathcal{L} = 0$

or

$$C_L = \frac{\mathcal{L}}{\frac{1}{2}\rho U^2 A} = \underline{\underline{0}}$$

Note:

F_1, F_2, F_3

$$\mathcal{D} = F_1 + (F_2 + F_3)\cos 45°$$

$$= \tfrac{1}{2}\left(\tfrac{1}{2}\rho U^2\right)\ell b + 2(1.2)\left(\tfrac{1}{2}\rho U^2\right)\frac{\ell b \cos 45°}{2\cos 45°}$$

$$= 1.7\left(\tfrac{1}{2}\rho U^2\right)\ell b$$

which checks with the above answer.

9.2R

9.2R (External flow character) A 0.23-m-diameter soccer ball moves through the air with a speed of 10 m/s. Would the flow around the ball be classified as low, moderate, or large Reynolds number flow? Explain.

(ANS: Large Reynolds number flow)

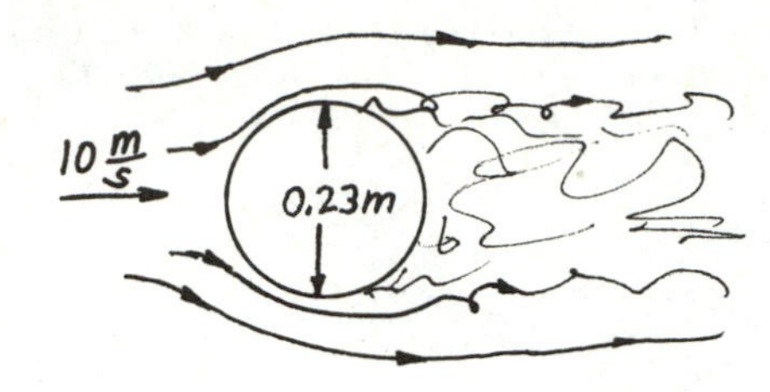

$Re = \frac{UD}{\nu}$, where $D = 0.23\,m$, $U = 10\,\frac{m}{s}$, and $\nu = 1.46 \times 10^{-5}\,\frac{m^2}{s}$ for standard air.

Thus,

$$Re = \frac{(10\,\frac{m}{s})(0.23\,m)}{1.46 \times 10^{-5}\,\frac{m^2}{s}} = 1.58 \times 10^5$$

This is a <u>large Reynolds number flow.</u>

(See Fig. 9.6c)

9.3R

9.3R (External flow character) A small 15-mm-long fish swims with a speed of 20 mm/s. Would a boundary layer type flow be developed along the sides of the fish? Explain.

(ANS: No)

$Re = \frac{U\ell}{\nu}$, or with $\ell = 15 \times 10^{-3}\,m$, $U = 20 \times 10^{-3}\,\frac{m}{s}$ and $\nu = 1.12 \times 10^{-6}\,\frac{m^2}{s}$ (i.e., 15.5 °C water)

$$Re = \frac{(20 \times 10^{-3}\,\frac{m}{s})(15 \times 10^{-3}\,m)}{1.12 \times 10^{-6}\,\frac{m^2}{s}} = 268$$

This Reynolds number is not large enough to have true boundary layer type flow. ($Re \approx 1000$ is often assumed to be the lower limit.)

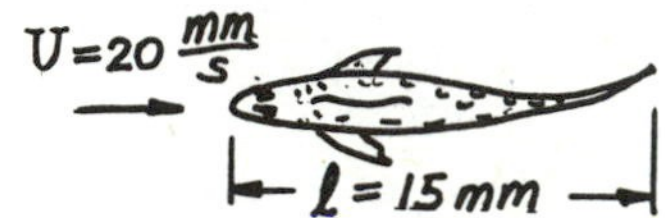

9.4R

9.4R (Boundary layer flow) Air flows over a flat plate of length $\ell = 2$ ft such that the Reynolds number based on the plate length is Re $= 2 \times 10^5$. Plot the boundary layer thickness, δ, for $0 \le x \le \ell$.

$$Re_\ell = \frac{U\ell}{\nu} = 2 \times 10^5, \text{ where } \ell = 2\text{ ft and } \nu = 1.57 \times 10^{-4} \frac{\text{ft}^2}{\text{s}}$$

Thus, $$U = \frac{(2\times10^5)(1.57\times10^{-4}\frac{\text{ft}^2}{\text{s}})}{2\text{ ft}} = 15.7 \frac{\text{ft}}{\text{s}}$$

For $Re_\ell \lesssim 5\times10^5$ the boundary layer flow is laminar and

$$\delta = 5\sqrt{\frac{\nu x}{U}} = 5\sqrt{\frac{(1.57\times10^{-4})x}{15.7}}, \text{ or } \delta = 0.0158\sqrt{x}, \text{ where } x \sim \text{ft}, \delta \sim \text{ft}$$

This result is plotted below for $0 \le x \le 2$ ft. Note the x-axis and δ-axis are different scale (the boundary layer is "thin").

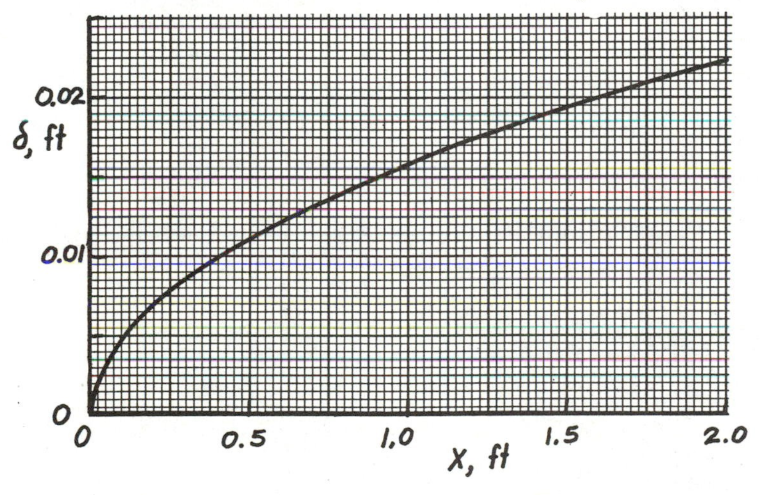

The boundary layer drawn to scale for $0 \le x \le 0.5$ ft is shown below. Note: It is "thin" (i.e., $\frac{\delta}{x} \ll 1$).

$U = 15.7 \frac{\text{ft}}{\text{s}}$

$u = 15.7 \frac{\text{ft}}{\text{s}}$

$u = 0$

Edge of boundary layer

$\delta\big|_{x=0.5\text{ft}} = 0.0112$ ft

δ

$x = 0$ $\quad x = 0.25$ ft $\quad x = 0.5$ ft

9.5R

9.5R (Boundary layer flow) At a given location along a flat plate the boundary layer thickness is $\delta = 45$ mm. At this location, what would be the boundary layer thickness if it were defined as the distance from the plate where the velocity is 97% of the upstream velocity rather than the standard 99%? Assume laminar flow.

(ANS: 38.5 mm)

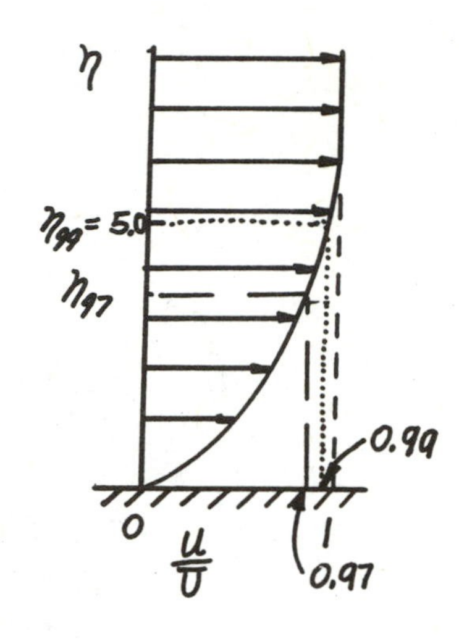

The standard definition of δ is the value of y where $u = 0.99\,U$. This occurs at a value of $\eta = \sqrt{\frac{U}{\nu x}}\,y = 5.0$ That is,

$$\delta_{99} = 45\text{ mm} \quad \text{when } \eta_{99} = 5.0$$

For given U, ν, and x

$$\frac{\eta_{97}}{\eta_{99}} = \frac{\delta_{97}(U/\nu x)^{1/2}}{\delta_{99}(U/\nu x)^{1/2}}$$

or

$$\delta_{97} = \delta_{99}\left(\frac{\eta_{97}}{\eta_{99}}\right) \qquad (1)$$

From Table 9.1 we interpolate to find η_{97}, the value of η when $\frac{u}{U} = 0.97$ Thus,

$$\eta_{97} = 4.0 + 0.4\left(\frac{0.9700 - 0.9555}{0.9759 - 0.9555}\right) = 4.28$$

so that Eq.(1) gives:

$$\delta_{97} = 45\text{ mm}\left(\frac{4.28}{5.00}\right) = \underline{\underline{38.5\text{ mm}}}$$

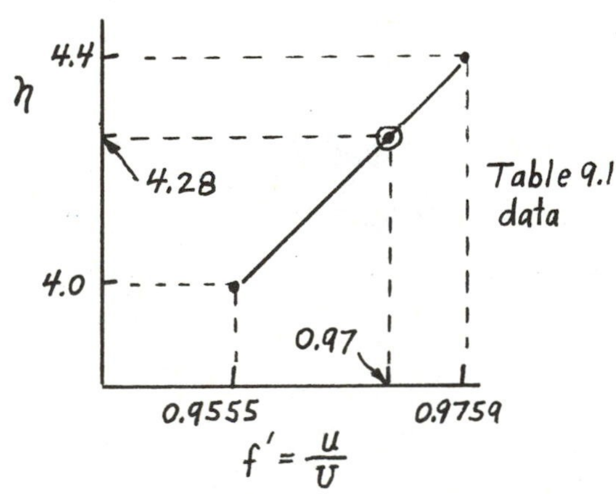

9.6R

9.6R (Friction drag) A laminar boundary layer formed on one side of a plate of length ℓ produces a drag $\mathscr{D}$. How much must the plate be shortened if the drag on the new plate is to be $\mathscr{D}/4$? Assume the upstream velocity remains the same. Explain your answer physically.

(ANS: $\ell_{new} = \ell/16$)

U → plate of length ℓ, drag $= \mathscr{D}_{f_1}$; U → plate of length ℓ_2, drag $= \mathscr{D}_{f_1}/4$

$$\mathscr{D}_f = \frac{1}{2}\rho U^2 C_{Df} A \quad \text{where} \quad C_{Df} = \frac{1.328}{\sqrt{Re_\ell}} = \frac{1.328}{\sqrt{\frac{U\ell}{\nu}}} \quad \text{and}$$

$A = b\ell$ where b = plate width.

Thus,

$$\mathscr{D}_f = \frac{1}{2}\rho U^2 \frac{1.328}{\sqrt{\frac{U\ell}{\nu}}} b\ell = 0.664\,\rho U^{\frac{3}{2}} b\sqrt{\nu}\sqrt{\ell} \qquad (1)$$

Consider two flows with $\rho_1 = \rho_2$, $U_1 = U_2$, $b_1 = b_2$, $\nu_1 = \nu_2$, and $\ell = \ell_1$ or ℓ_2 so from Eq. (1)

$$\frac{\mathscr{D}_{f1}}{\mathscr{D}_{f2}} = \sqrt{\frac{\ell_1}{\ell_2}} \quad \text{so that with } \mathscr{D}_{f_2} = \frac{1}{4}\mathscr{D}_{f_1}, \quad \sqrt{\frac{\ell}{\ell_2}} = 4 \quad \text{or} \quad \ell_2 = \frac{\ell}{16}$$

Since the boundary layer is thinner on the front portion of the plate than on the rear portion, the wall shear stress is greater near the front. To reduce the drag by a factor of four, more than three-fourths of the plate must be removed (i.e., from length ℓ to length $\frac{\ell}{16}$).

9.7R

9.7R (Momentum integral equation) As is indicated in Table 9.2, the laminar boundary layer results obtained from the momentum integral equation are relatively insensitive to the shape of the assumed velocity profile. Consider the profile given by $u = U$ for $y > \delta$, and $u = U\{1 - [(y - \delta)/\delta]^2\}^{1/2}$ for $y \le \delta$ as shown in Fig. P9.7R. Note that this satisfies the conditions $u = 0$ at $y = 0$ and $u = U$ at $y = \delta$. However, show that such a profile produces meaningless results when used with the momentum integral equation. Explain.

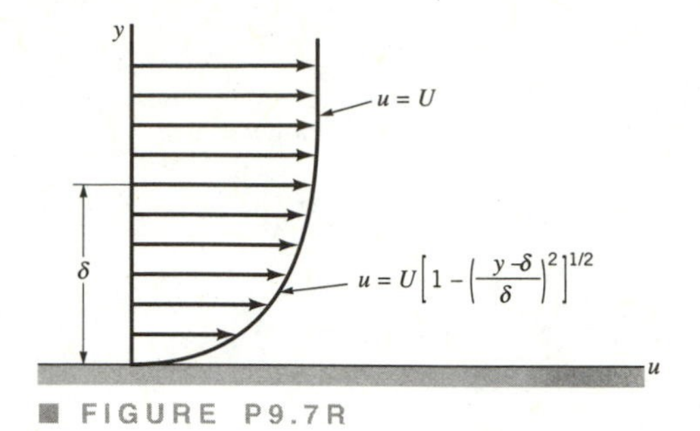

■ FIGURE P9.7R

From the momentum integral equation,

$$\delta = \sqrt{\frac{2C_2 \nu x}{U C_1}}, \quad \text{where } \frac{u}{U} = g(Y) = \left[1 - (Y-1)^2\right]^{\frac{1}{2}} \text{ with } Y = \frac{y}{\delta}. \qquad (1)$$

Note: $\frac{u}{U} = 0$ at $Y = 0$ and $\frac{u}{U} = 1$ and $Y = 1$, as required.

Also, $C_1 = \int_0^1 g(1-g)\,dY$ which can be evaluated for the given $g(Y)$.

However,

$$C_2 = \left.\frac{dg}{dY}\right|_{Y=0}, \quad \text{where } \frac{dg}{dY} = \frac{1}{2}\left[1-(Y-1)^2\right]^{-\frac{1}{2}}(-2)(Y-1) = \frac{(1-Y)}{\left[1-(Y-1)^2\right]^{\frac{1}{2}}}$$

Thus, as $Y \to 0$, $\frac{dg}{dY} \to \infty$ so that

$C_2 = \infty$, which from Eq.(1) gives $\delta = \infty$

This profile cannot be used since it gives $\delta = \infty$ due to the physically unrealistic $\frac{\partial u}{\partial y} = \infty$ at the surface ($y = 0$).

See the figure below.

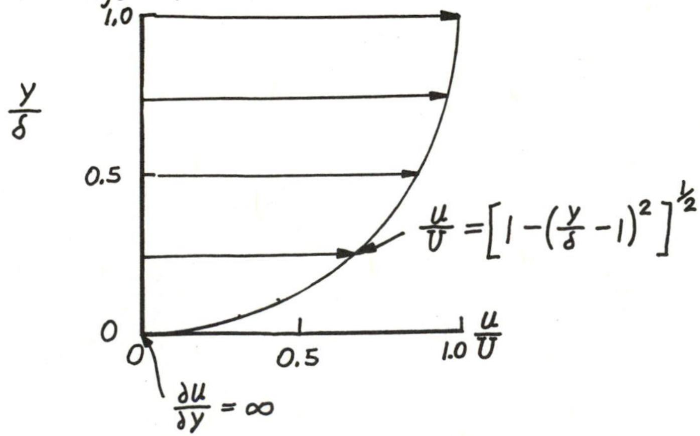

9.8R

9.8R (Drag—low Reynolds number) How fast do small water droplets of 0.06-μm (6×10^{-8} m) diameter fall through the air under standard sea-level conditions? Assume the drops do not evaporate. Repeat the problem for standard conditions at 5000-m altitude.

(ANS: 1.10×10^{-7} m/s; 1.20×10^{-7} m/s)

dia. $= D = 6 \times 10^{-8}$ m

$Re = \frac{UD}{\nu}$

For steady conditions, $\mathcal{D} + F_B = W$,
where, if $Re = \frac{UD}{\nu} < 1$,

$\mathcal{D} = \text{drag} = 3\pi D U \mu$ Also, $W = \gamma_{H_2O} \forall = \gamma_{H_2O} \frac{4}{3}\pi\left(\frac{D}{2}\right)^3 = \text{weight}$

and $F_B = \gamma_{air} \forall = \gamma_{air} \frac{4}{3}\pi\left(\frac{D}{2}\right)^3 = \text{buoyant force}$

Since $\gamma_{air} \ll \gamma_{H_2O}$, we can neglect the buoyant force.
That is, $\mathcal{D} = W$, or

$$3\pi D U \mu = \gamma_{H_2O} \frac{4\pi}{3}\left(\frac{D}{2}\right)^3, \text{ or } U = \frac{\gamma_{H_2O} D^2}{18\mu} \qquad (1)$$

At sea level $\mu = 1.789 \times 10^{-5} \frac{N \cdot s}{m^2}$ (see Table C.2) so that

$$U = \frac{(9.80 \times 10^3 \frac{N}{m^3})(6 \times 10^{-8} m)^2}{18(1.789 \times 10^{-5} \frac{N \cdot s}{m^2})} = \underline{\underline{1.10 \times 10^{-7} \frac{m}{s}}}$$

Note that $Re = \frac{(1.10 \times 10^{-7} \frac{m}{s})(6 \times 10^{-8} m)}{1.46 \times 10^{-5} \frac{m^2}{s}} = 4.52 \times 10^{-10} \ll 1$ so the use of the low Re drag equation is valid.

At an altitude of 5000 m, $\mu = 1.628 \times 10^{-5} \frac{N \cdot s}{m^2}$ and from Eq.(1)

$$U = \frac{(9.80 \times 10^3 \frac{N}{m^3})(6 \times 10^{-8} m)^2}{18(1.628 \times 10^{-5} \frac{N \cdot s}{m^2})} = \underline{\underline{1.20 \times 10^{-7} \frac{m}{s}}}$$

9.9R

9.9R (Drag) A 12-mm-diameter cable is strung between a series of poles that are 40 m apart. Determine the horizontal force this cable puts on each pole if the wind velocity is 30 m/s.
(ANS: 372 N)

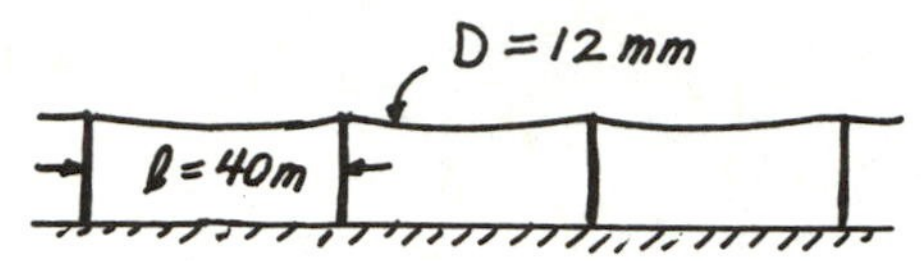

F_p = force on one pole = $\mathcal{D}$

where $\mathcal{D} = C_D \frac{1}{2} \rho U^2 A$

Since $Re = \frac{UD}{\nu} = \frac{(30\frac{m}{s})(0.012m)}{1.46\times10^{-5}\frac{m^2}{s}} = 2.47\times10^4$ it follows from Fig. 9.21 that

$C_D = 1.4$. Hence, $F_p = 1.4(\frac{1}{2})(1.23\frac{kg}{m^3})(30\frac{m}{s})^2(40\,m)(0.012m) = \underline{\underline{372\,N}}$

9.10R

9.10R (Drag) How much less power is required to pedal a racing-style bicycle at 20 mph with a 10-mph tail wind than at the same speed with a 10-mph head wind? (See Fig. 9.30.)
(ANS: 0.375 hp)

$U = (20+10)\ mph$
$= 30\ mph$

head wind

$\mathcal{P}$ = power to overcome aerodynamic drag = $U_B \mathcal{D}$, where U_B = bike speed and $\mathcal{D} = \frac{1}{2}\rho U^2 C_D A$, with U = wind speed relative to bike.

From Fig. 9.30, for a racing bike $C_D = 0.88$ and $A = 3.9\ ft^2$

With a tail wind $U_B = (20\ mph)\left(\frac{88\ \frac{ft}{s}}{60\ mph}\right) = 29.3\ \frac{ft}{s}$

and $U = (20-10)\ mph = 14.7\ \frac{ft}{s}$

Thus, $\mathcal{P}_{tail} = (29.3\ \frac{ft}{s})\frac{1}{2}(0.00238\ \frac{slugs}{ft^3})(14.7\ \frac{ft}{s})^2(3.9\ ft^2)(0.88)$

$= (25.9\ \frac{ft \cdot lb}{s})\left(\frac{1\ hp}{550\ \frac{ft \cdot lb}{s}}\right) = 0.0471\ hp$

With a head wind $U = (20+10)\ mph = 44\ \frac{ft}{s}$

Thus, $\mathcal{P}_{head} = (29.3\ \frac{ft}{s})\frac{1}{2}(0.00238\ \frac{slugs}{ft^3})(44\ \frac{ft}{s})^2(3.9\ ft^2)(0.88)$

$= (232\ \frac{ft \cdot lb}{s})\left(\frac{1\ hp}{550\ \frac{ft \cdot lb}{s}}\right) = 0.422\ hp$

Hence,

$\mathcal{P}_{head} - \mathcal{P}_{tail} = (0.422 - 0.0471)\ hp = \underline{\underline{0.375\ hp}}$

9.11R

9.11R (Drag) A rectangular car-top carrier of 1.6-ft height, 5.0-ft length (front to back), and a 4.2-ft width is attached to the top of a car. Estimate the additional power required to drive the car with the carrier at 60 mph through still air compared with the power required to drive only the car at 60 mph.

(ANS: 12.9 hp)

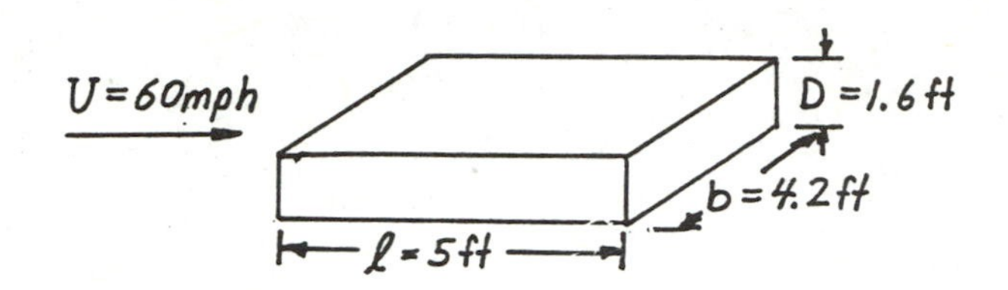

$\mathcal{P}$ = power = $U\mathcal{D}$ where $\mathcal{D}$ = drag = $C_D \frac{1}{2} \rho U^2 A$ (1)

with $U = \left(60 \frac{mi}{hr}\right)\left(\frac{1\ hr}{3600\ s}\right)\left(\frac{5280\ ft}{mi}\right) = 88 \frac{ft}{s}$

From Fig. 9.28 with $\frac{\ell}{D} = \frac{5}{1.6} = 3.13$ we obtain $C_D = 1.3$

(Note: $Re = \frac{UD}{\nu} = \frac{(88 \frac{ft}{s})(1.6\ ft)}{1.57 \times 10^{-4} \frac{ft^2}{s}} = 8.97 \times 10^5$, whereas the C_D value from Fig. 9.28 is that for $Re = 10^5$. However, for a blunt object like this box, the value of C_D is essentially independent of Re.)

Thus, from Eq. (1)

$\mathcal{D} = 1.3\left(\frac{1}{2}\right)\left(0.00238 \frac{slugs}{ft^3}\right)(1.6\ ft)(4.2\ ft)\left(88 \frac{ft}{s}\right)^2 = 80.5\ lb$

so that

$\mathcal{P} = U\mathcal{D} = \left(88 \frac{ft}{s}\right)(80.5\ lb) \frac{1\ hp}{550 \frac{ft \cdot lb}{s}} = \underline{\underline{12.9\ hp}}$

9.12R

9.12R (Drag) Estimate the wind velocity necessary to blow over the 250-kN boxcar shown in Fig. P9.12R.
(ANS: approximately 32.6 m/s to 35.1 m/s)

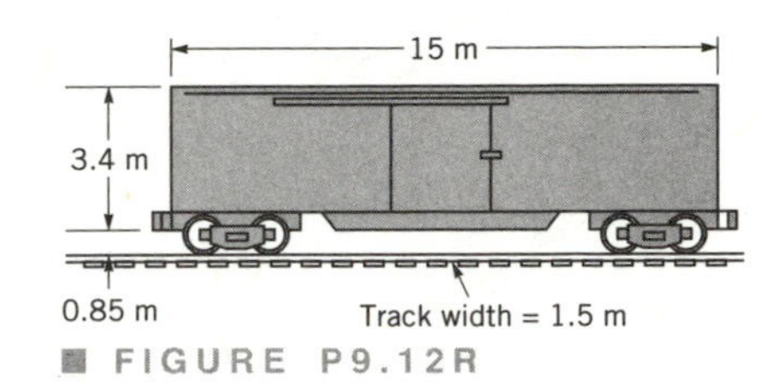

■ FIGURE P9.12R

If the boxcar is about to tip around point O, then $\Sigma M_0 = 0$, or

$$\left(\frac{3.4}{2} + 0.85\right) m = 2.55 m$$

$$2.55\,\mathscr{D} = 0.75\,W$$

Hence,

$$\mathscr{D} = \frac{0.75(250 \times 10^3 N)}{2.55} = 7.34 \times 10^4 N,$$

where

$$\mathscr{D} = C_D \tfrac{1}{2} \rho U^2 A \quad \text{or} \quad U = \sqrt{\frac{2\mathscr{D}}{C_D \rho A}}$$

If we consider the boxcar as a flat plate, then from Fig. 9.19 $C_D = 1.9$ so that

$$U = \left[\frac{2\,(7.34 \times 10^4 N)}{1.9\,(1.23 \frac{kg}{m^3})(3.4m)(15m)}\right]^{1/2} = 35.1 \frac{m}{s}$$

Note: The C_D value used above is for a thin plate normal to the flow (i.e. $\ell = 0$ in the above figure). From Fig. 9.28 with $\ell/D = 1$ (i.e. assume the width equals the height) we have $C_D = 2.2$ which gives $U = 32.6 \frac{m}{s}$. The two values are not too different.

9.13R

9.13R (Drag) A 200-N rock (roughly spherical in shape) of specific gravity $SG = 1.93$ falls at a constant speed U. Determine U if the rock falls through **(a)** air; **(b)** water.

(ANS: 176 m/s; 5.28 m/s)

F_B, γ_r, W, dia. D, μ, ρ, $\mathcal{D}$, U, z, x

For steady fall $\sum F_z = 0$ or $F_B + \mathcal{D} = W$ (1)

where $\mathcal{D} = C_D \frac{1}{2}\rho U^2 \frac{\pi}{4} D^2$ and $W = \gamma_r \forall = \gamma_r \frac{4\pi}{3}\left(\frac{D}{2}\right)^3$ $\quad \gamma_r = SG\,\gamma_{H_2O}$

Hence, $W = 200\,N = 1.93\left(9.80\times 10^3 \frac{N}{m^3}\right)\frac{4\pi}{3}\left(\frac{D}{2}\right)^3$, or $D = 0.272\,m$

Also, $F_B = \rho g \forall = \rho\left(9.81 \frac{m}{s^2}\right)\frac{4\pi}{3}\left(\frac{0.272}{2}m\right)^3 = 0.103\,\rho\ N$, where $\rho \sim \frac{kg}{m^3}$

Thus, Eq.(1) becomes

$0.103\,\rho + C_D \frac{1}{2}\rho U^2 \frac{\pi}{4}(0.272)^2 = 200$, or $0.103\,\rho + 0.0291\,\rho C_D U^2 = 200$, where $U \sim \frac{m}{s}$ (2)

a) For falling through air, $\rho = 1.23 \frac{kg}{m^3}$ and Eq.(2) becomes

$0.0358\, C_D U^2 = 199.9$ or $C_D U^2 = 5580$ (3)

Also, $Re = \frac{UD}{\nu} = \frac{(0.272\,m)\,U}{1.46\times 10^{-5}\frac{m^2}{s}} = 1.86\times 10^4\, U$, where $U \sim \frac{m}{s}$ (4)

and from Fig. 9.25, C_D vs. Re — smooth sphere (5)

Note: Is the rock a smooth sphere?

Trial and error solution: Assume C_D; obtain U from Eq.(3), Re from Eq.(4); check C_D from Eq.(5), the graph.

Assume $C_D = 0.5 \rightarrow U = 106 \frac{m}{s} \rightarrow Re = 1.97\times 10^6 \rightarrow C_D = 0.16 \neq 0.5$

Assume $C_D = 0.16 \rightarrow U = 187 \frac{m}{s} \rightarrow Re = 3.47\times 10^6 \rightarrow C_D = 0.18 \neq 0.16$

Assume $C_D = 0.18 \rightarrow U = 176 \frac{m}{s} \rightarrow Re = 3.27\times 10^6 \rightarrow C_D = 0.18$ (checks)

Thus, $U \approx 176 \frac{m}{s}$ in air.

(b) For falling through water, $\rho = 999 \frac{kg}{m^3}$ and Eq.(2) becomes

$C_D U^2 = 3.34$ (6)

Also, $Re = \frac{UD}{\nu} = \frac{(0.272\,m)\,U}{1.12\times 10^{-6}\frac{m^2}{s}} = 2.43\times 10^5\, U$ (7)

Trial and error solution with Eqs. (6), (7), and (5).

Assume $C_D = 0.2 \rightarrow U = 4.09 \frac{m}{s} \rightarrow Re = 9.94\times 10^5 \rightarrow C_D = 0.10 \neq 0.2$

Assume $C_D = 0.12 \rightarrow U = 5.28 \frac{m}{s} \rightarrow Re = 1.28\times 10^6 \rightarrow C_D = 0.12$ (checks)

Thus, $U \approx 5.28 \frac{m}{s}$ in water.

9.14R

9.14R (Drag—composite body) A shortwave radio antenna is constructed from circular tubing, as is illustrated in Fig. P9.14R. Estimate the wind force on the antenna in a 100 km/hr wind.

(ANS: 180 N)

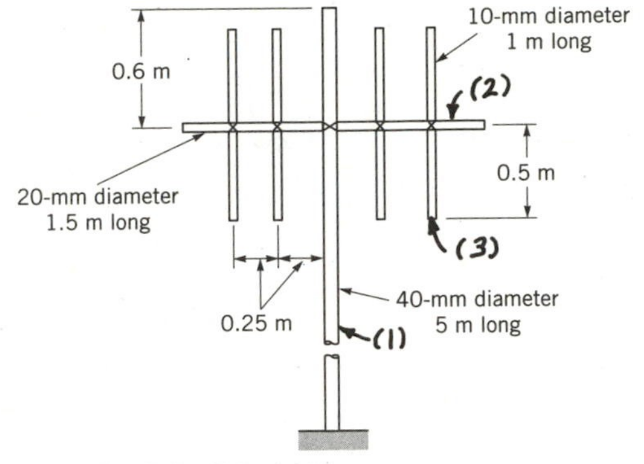

■ FIGURE P9.14R

The antenna is a composite body consisting of one main pole, one horizontal bar, and four vertical rods. Thus,

$$\mathcal{D} = \mathcal{D}_1 + \mathcal{D}_2 + 4\mathcal{D}_3$$

$$= \frac{1}{2}\rho U^2 \left[C_{D_1} A_1 + C_{D_2} A_2 + C_{D_3} A_3\right]$$

where $U = 100 \frac{km}{hr}\left(\frac{10^3 m}{km}\right)\left(\frac{1\,h}{3600\,s}\right) = 27.8 \frac{m}{s}$

Obtain C_{D_i} from Fig. 9.21 for the given $Re_i = \frac{UD_i}{\nu}$.

Thus, $Re_1 = \dfrac{(27.8 \frac{m}{s})(0.04 m)}{1.46 \times 10^{-5} \frac{m^2}{s}} = 7.62 \times 10^4 \longrightarrow C_{D_1} = 1.4$

$Re_2 = \dfrac{(27.8 \frac{m}{s})(0.02 m)}{1.46 \times 10^{-5} \frac{m^2}{s}} = 3.81 \times 10^4 \longrightarrow C_{D_2} = 1.4$

and

$Re_3 = \dfrac{(27.8 \frac{m}{s})(0.01 m)}{1.46 \times 10^{-5} \frac{m^2}{s}} = 1.90 \times 10^4 \longrightarrow C_{D_3} = 1.4 = C_{D_2} = C_{D_1}$

so that

$$\mathcal{D} = \frac{1}{2}\left(1.23 \frac{kg}{m^3}\right)\left(27.8 \frac{m}{s}\right)^2 (1.4)\left[(5 m)(0.04 m) + (1.5 m)(0.02 m) + 4(1 m)(0.01 m)\right]$$

or

$$\mathcal{D} = \underline{\underline{180\ N}}$$

9.15R

9.15R (Lift) Show that for level flight the drag on a given airplane is independent of altitude if the lift and drag coefficients remain constant. Note that with C_L constant the airplane must fly faster at a higher altitude.

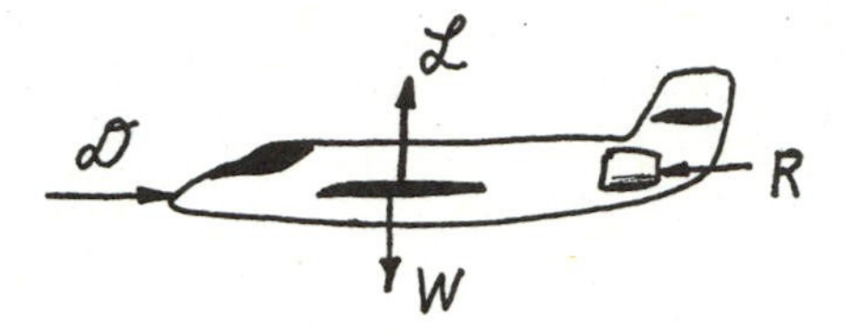

For level flight $\mathscr{L} = W$, where W = airplane weight = constant

and $\mathscr{L} = C_L \frac{1}{2}\rho U^2 A$ (1)

Also, $\mathscr{D} = C_D \frac{1}{2}\rho U^2 A$ (2)

Thus, if C_L, A, and $\mathscr{L}$ are constant (independent of altitude), it follows from Eq. (1) that ρU^2 is also constant. Hence, for constant C_D it follows from Eq. (2) that $\mathscr{D}$ is constant (independent of altitude). (Since ρ decreases with increasing altitude the idea that ρU^2 is constant implies that U increases with altitude.)

9.16R

9.16R (Lift) The wing area of a small airplane weighing 6.22 kN is 10.2 m^2. **(a)** If the cruising speed of the plane is 210 km/hr, determine the lift coefficient of the wing. **(b)** If the engine delivers 150 kW at this speed, and if 60% of this power represents propeller loss and body resistance, what is the drag coefficient of the wing.

(ANS: 0.292; 0.0483)

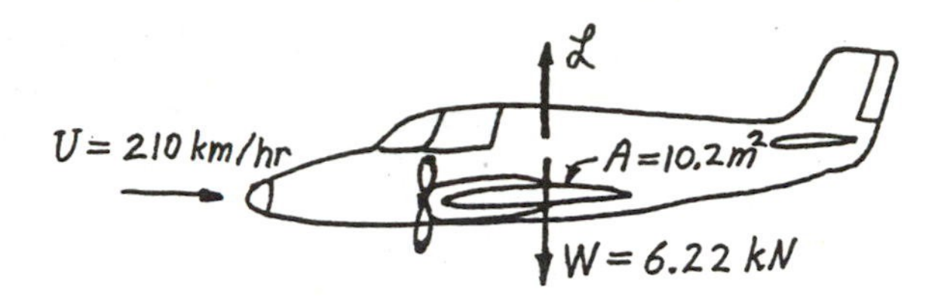

For equilibrium, $\mathcal{L} = W = 6.22\ kN$, where $\mathcal{L} = C_L \frac{1}{2}\rho U^2 A$.

Thus, with $U = (210\ \frac{km}{hr})(\frac{1\ hr}{3600\ s})(\frac{1000\ m}{1\ km}) = 58.3\ \frac{m}{s}$

$$C_L = \frac{\mathcal{L}}{\frac{1}{2}\rho U^2 A} = \frac{6.22 \times 10^3\ N}{\frac{1}{2}(1.23\ \frac{kg}{m^3})(58.3\ \frac{m}{s})^2(10.2\ m^2)} = \underline{\underline{0.292}}$$

Also,

$$\mathcal{P}_{total} = \mathcal{P}_{body} + \mathcal{P}_{\substack{loss\\prop}} + \mathcal{P}_{wing}, \text{ where } \mathcal{P}_{body} + \mathcal{P}_{\substack{loss\\prop}} = 0.6\ \mathcal{P}_{total} = 0.6(150\ kW) = 90\ kW$$

Thus,

$$\mathcal{P}_{wing} = 150\ kW - 90\ kW = 60\ kW$$

where

$$\mathcal{P}_{wing} = \mathcal{D}U = C_D \frac{1}{2}\rho U^2 A U$$

or

$$C_D = \frac{2\mathcal{P}_{wing}}{\rho U^3 A} = \frac{2\ (60 \times 10^3\ \frac{N \cdot m}{s})}{(1.23\ \frac{kg}{m^3})(58.3\ \frac{m}{s})^3(10.2\ m^2)} = \underline{\underline{0.0483}}$$

10
Open-Channel Flow

Motion of water induced by surface waves: As a wave passes along the surface of the water, the water particles follow elliptical paths. There is no net motion of the water, just a periodic, cyclic trajectory (neutrally buoyant particles in water). (Photograph by A. Wallet and F. Ruellan, Ref. 13, courtesy of M. C. Vasseur, Sogreah.)

10.1R

10.1R (Surface waves) If the water depth in a pond is 15 ft, determine the speed of small amplitude, long wavelength ($\lambda \gg y$) waves on the surface.

(ANS: 22.0 ft/s)

For $\lambda \gg y$, $\tanh \frac{2\pi y}{\lambda} \approx \frac{2\pi y}{\lambda}$ so that (see Eq 10.4)

$$c = \left[\frac{g\lambda}{2\pi}\tanh\left(\frac{2\pi y}{\lambda}\right)\right]^{\frac{1}{2}} \approx \left[\frac{g\lambda}{2\pi}\left(\frac{2\pi y}{\lambda}\right)\right]^{\frac{1}{2}} = \sqrt{gy}$$

or

$$c = \left[\left(32.2\,\frac{ft}{s^2}\right)(15\,ft)\right]^{\frac{1}{2}} = \underline{\underline{22.0\,\frac{ft}{s}}}$$

10.2R

10.2R (Surface waves) A small amplitude wave travels across a pond with a speed of 9.6 ft/s. Determine the water depth.

(ANS: $y \geq 2.86$ ft)

For long wavelength waves $c = \sqrt{gy}$ or $y = \frac{c^2}{g} = \frac{(9.6\,\frac{ft}{s})^2}{32.2\,\frac{ft}{s^2}} = 2.86\,ft$

In general, from Eq. 10.4:

$$c = \left[\frac{g\lambda}{2\pi}\tanh\left(\frac{2\pi y}{\lambda}\right)\right]^{\frac{1}{2}}, \text{ or } 9.6\,\frac{ft}{s} = \left[\frac{(32.2\,\frac{ft}{s^2})\lambda}{2\pi}\tanh\left(\frac{2\pi y}{\lambda}\right)\right]^{\frac{1}{2}}$$

Hence: $\tanh\left(\frac{2\pi y}{\lambda}\right) = \frac{18.0}{\lambda}$ Pick various λ and calculate y. (1)

Note: Since $\tanh\left(\frac{2\pi y}{\lambda}\right) \leq 1$, it is not possible to have $\lambda < 18$ ft (for $c = 9.6$ ft/s). Also, as $\lambda \to \infty$, $y \to 2.86$ ft (the long wavelength limit).

λ, ft	y, ft
18	∞
19	5.46
20	4.69
25	3.61
30	3.31
50	3.00
100	2.90
∞	2.86

Although we must know the wavelength, λ, to determine the depth, y, we do know that $y \geq \underline{2.86\text{ ft}}$

10.3R

10.3R (Froude number) The average velocity and average depth of a river from its beginning in the mountains to its discharge into the ocean are given in the table below. Plot a graph of the Froude number as a function of distance along the river.

Distance (mi)	Average Velocity (ft/s)	Average Depth (ft)
0	13	1.5
5	10	2.0
10	9	2.3
30	5	3.7
50	4	5.4
80	4	6.0
90	3	6.2

(ANS: Fr = 1.87 at the beginning, 0.212 at the discharge)

$Fr = \frac{V}{c} = \frac{V}{\sqrt{g\,y}}$, where $g = 32.2\ \frac{ft}{s^2}$ and values of V and y are given in the table. The following results are obtained.

Distance, mi	Fr
0	1.87
5	1.25
10	1.05
30	0.458
50	0.303
80	0.288
90	0.212

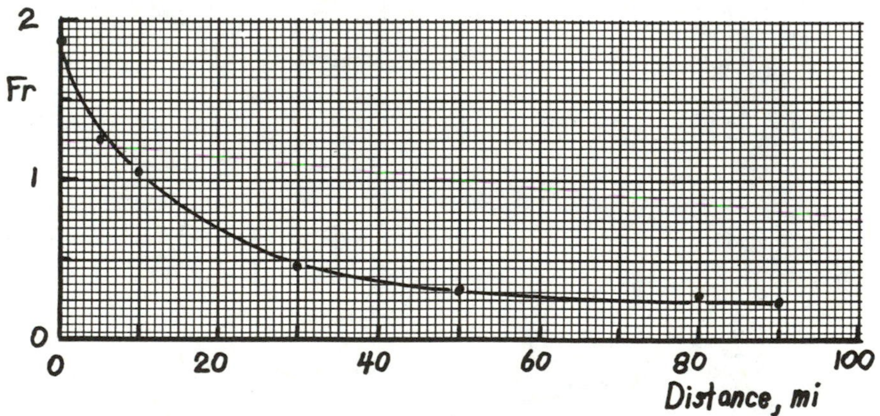

Note that the flow is supercritical for the first 10 miles or so.

10.4R

10.4R (Froude number) Water flows in a rectangular channel at a depth of 4 ft and a flowrate of $Q = 200$ cfs. Determine the minimum channel width if the flow is to be subcritical.

(ANS: 4.41 ft)

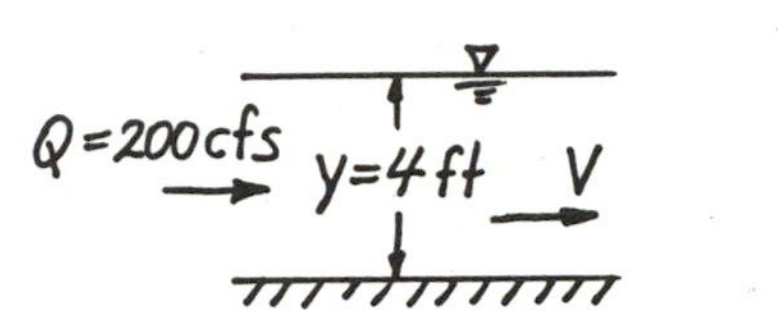

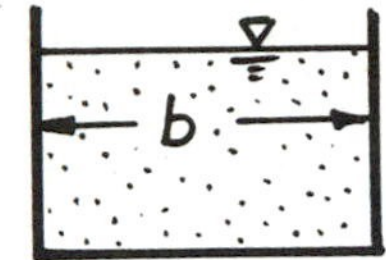

$$V = \frac{Q}{A} = \frac{200\,\frac{ft^3}{s}}{(4ft)\,b} = \frac{50}{b}, \text{ where } b = \text{width} \sim ft \text{ and } V \sim ft/s.$$

Thus, $$Fr = \frac{V}{\sqrt{gy}} = \frac{\left(\frac{50}{b}\right)}{\left[(32.2\,\frac{ft}{s^2})(4ft)\right]^{1/2}}$$ Note: As b decreases, Fr increases.

Set $Fr = 1$ for minimum width for subcritical flow.

Hence,

$$1 = \frac{50/b}{\left[(32.2)(4)\right]^{1/2}} \quad \text{or} \quad b = \underline{\underline{4.41\,ft}}$$

10.5R

10.5R (Specific energy) Plot the specific energy diagram for a wide channel carrying $q = 50 \text{ ft}^2/\text{s}$. Determine **(a)** the critical depth, **(b)** the minimum specific energy, **(c)** the alternate depth corresponding to a depth of 2.5 ft, and **(d)** the possible flow velocities if $E = 10$ ft.
(ANS: 4.27 ft; 6.41 ft; 8.12 ft; 5.22 ft/s or 22.3 ft/s)

$$E = y + \frac{q^2}{2g y^2}, \text{ or } E = y + \frac{(50 \frac{ft^2}{s})^2}{2(32.2 \frac{ft}{s^2}) y^2}$$

Thus,

$$E = y + \frac{38.8}{y^2}, \text{ where } E \sim ft,\ y \sim ft \quad \text{Equation (1) is plotted below.} \quad (1)$$

Note:

a) $y_c = \left(\frac{q^2}{g}\right)^{\frac{1}{3}} = \left(\frac{(50 \frac{ft^2}{s})^2}{(32.2 \frac{ft}{s^2})}\right)^{\frac{1}{3}} = \underline{\underline{4.27 \text{ ft}}}$

b) $E_{min} = \frac{3}{2} y_c = \frac{3}{2}(4.27 \text{ ft}) = \underline{\underline{6.41 \text{ ft}}}$

c) If $y = 2.5$ ft, then $E = 2.5 + \frac{38.8}{(2.5)^2} = 8.71$

Thus, solve for the positive real roots of

$$8.71 = y + \frac{38.8}{y^2}, \text{ or } y^3 - 8.71 y^2 + 38.8 = 0 \quad (2)$$

One root is the given $y = 2.5$. Thus, divide $(y - 2.5)$ into Eq. (2) to obtain $y^2 - 6.21y - 15.53 = 0$ which has roots

$$y = \frac{6.21 \pm \sqrt{6.21^2 + 4(15.53)}}{2} = 8.12 \text{ ft, or } y = -1.91 \text{ ft}$$

Thus, the alternate depth corresponding to $y = 2.5$ ft is $y = \underline{\underline{8.12 \text{ ft}}}$

d) For $E = 10$ ft Eq. (1) is $10 = y + \frac{38.8}{y^2}$, or $y^3 - 10y^2 + 38.8 = 0 \quad (3)$

which has roots: $y = 9.57, 2.24,$ and -1.81. Thus, $y = 9.57$ ft or $y = 2.24$ ft

The corresponding velocities are $V = \frac{q}{y} = \frac{50 \frac{ft^2}{s}}{9.57 \text{ ft}} = \underline{\underline{5.22 \frac{ft}{s}}}$

$$V = \frac{50 \frac{ft^2}{s}}{2.24 \text{ ft}} = \underline{\underline{22.3 \frac{ft}{s}}}$$

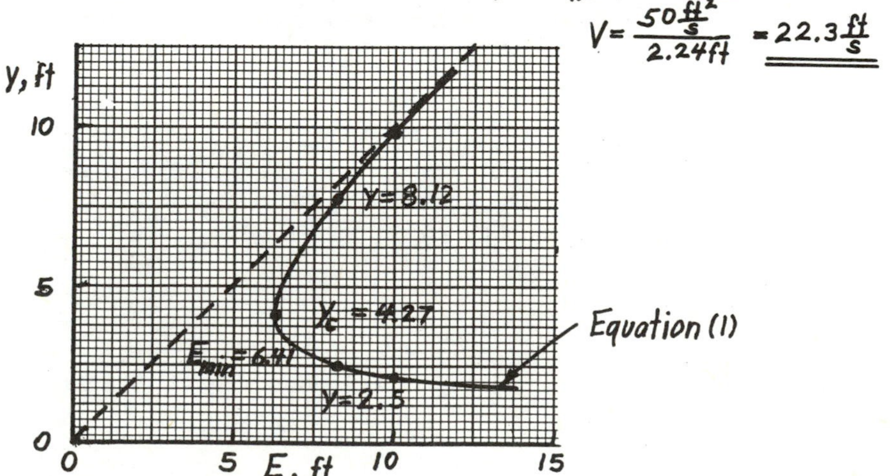

The specific energy diagram is as shown above.

10.6R

10.6R (Specific energy) Water flows at a rate of 1000 ft^3/s in a horizontal rectangular channel 30 ft wide with a 2-ft depth. Determine the depth if the channel contracts to a width of 25 ft. Explain.

(ANS: 2.57 ft)

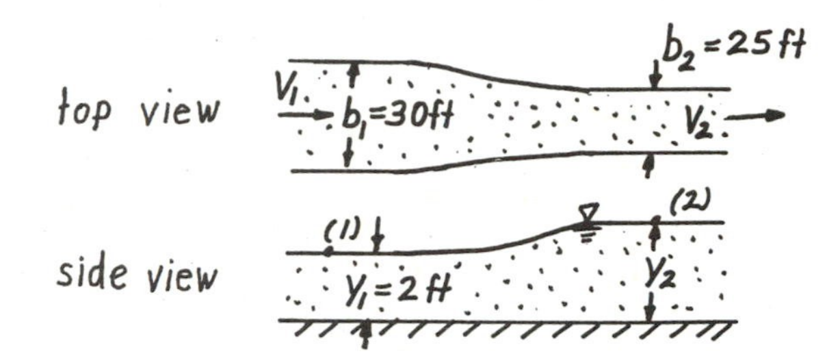

$$\frac{p_1}{\gamma} + \frac{V_1^2}{2g} + z_1 = \frac{p_2}{\gamma} + \frac{V_2^2}{2g} + z_2, \text{ where } z_1 = y_1 = 2\text{ ft},\ z_2 = y_2,\ p_1 = p_2 = 0,$$

$$V_1 = \frac{Q}{A_1} = \frac{1000\frac{ft^3}{s}}{(2ft)(30ft)} = 16.7\frac{ft}{s}, \text{ and } V_2 = \frac{Q}{A_2} = \frac{1000\frac{ft^3}{s}}{(25ft)y_2} = \frac{40}{y_2}\frac{ft}{s} \text{ with } y_2 \sim ft$$

Thus,

$$\frac{(16.7\frac{ft}{s})^2}{2(32.2\frac{ft}{s^2})} + 2\text{ ft} = \frac{(\frac{40}{y_2})^2}{2(32.2\frac{ft}{s^2})} + y_2$$

or

$$y_2^3 - 6.33y_2^2 + 24.8 = 0 \text{ which has roots } y_2 = 5.51,\ 2.57,\ -1.75\text{ ft.}$$

Note: $Fr_1 = \frac{V_1}{\sqrt{gy_1}} = \frac{16.7\frac{ft}{s}}{[(32.2\frac{ft}{s})(2ft)]^{\frac{1}{2}}} = 2.08 > 1$

If there is no relative minimum area between (1) and (2) where critical flow can occur, it follows that $Fr_2 > 1$ also. Thus, the $y_2 = 5.51$ (subcritical flow root) is not valid.

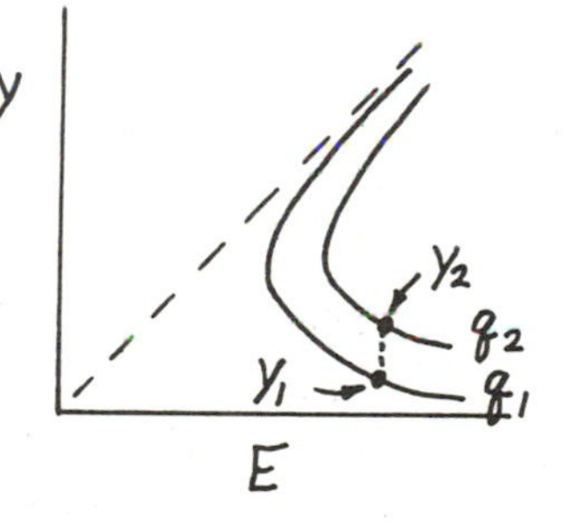

Thus, $y_2 = z_2 = \underline{\underline{2.57\text{ ft}}}$

Note: If $y_2 = 5.51$ ft and $V_2 = \frac{40}{5.51} = 7.26\frac{ft}{s}$, then

$$Fr_2 = \frac{V_2}{\sqrt{gy_2}} = \frac{7.26\text{ ft/s}}{[(32.2\text{ ft/s}^2)(5.51\text{ ft})]^{\frac{1}{2}}} = 0.545 < 1,$$

whereas if

$$y_1 = 2.57\text{ft and } V_2 = \frac{40}{2.57} = 15.6\frac{ft}{s}, \text{ then}$$

$$Fr_2 = \frac{15.6}{[(32.2)(2.57)]^{\frac{1}{2}}} = 1.71 > 1$$

10.7R

10.7R (Wall shear stress) Water flows in a 10-ft-wide rectangular channel with a flowrate of 150 cfs and a depth of 3 ft. If the slope is 0.005, determine the Manning coefficient, n, and the average shear stress at the sides and bottom of the channel. (ANS: 0.0320; 0.585 lb/ft²)

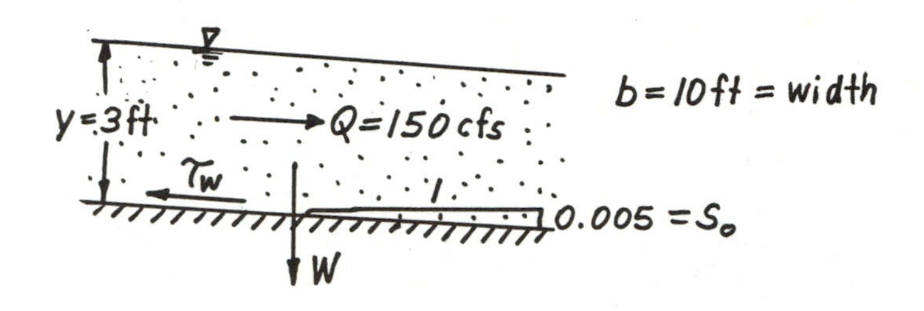

$$Q = \frac{\kappa}{n} A R_h^{2/3} S_o^{1/2}, \text{ where } \kappa = 1.49,\ A = by = (10\,ft)(3\,ft) = 30\,ft^2$$

$$P = b + 2y = 10\,ft + 2(3\,ft) = 16\,ft, \text{ or } R_h = \frac{A}{P} = \frac{30\,ft^2}{16\,ft} = 1.875\,ft$$

Thus,

$$n = \frac{1.49 A R_h^{2/3} S_o^{1/2}}{Q} = \frac{1.49(30)(1.875)^{2/3}(0.005)^{1/2}}{150} = \underline{\underline{0.0320}}$$

Also,

$$\tau_w = \gamma R_h S_o = \left(62.4\,\frac{lb}{ft^3}\right)(1.875\,ft)(0.005) = \underline{\underline{0.585\,\frac{lb}{ft^2}}}$$

10.8R

10.8R (Manning equation) The triangular flume shown in Fig. P10.8R is built to carry its design flowrate, Q_0, at a depth of 0.90 m as is indicated. If the flume is to be able to carry up to twice its design flowrate, $Q = 2Q_0$, determine the freeboard, ℓ, needed.

(ANS: 0.378 m)

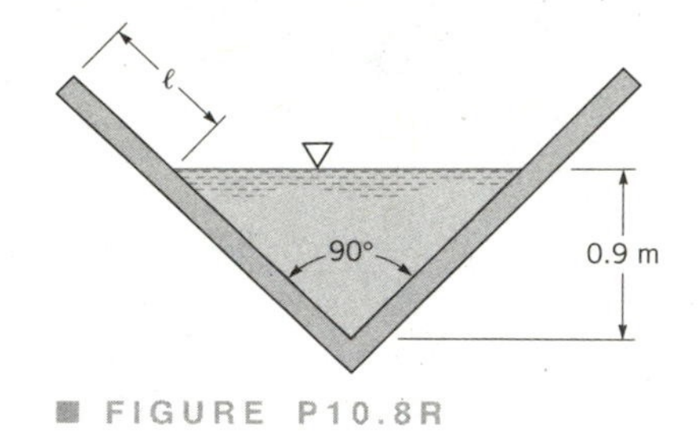

■ FIGURE P10.8R

Let $(\;)_0$ denote the design conditions and $(\;)_2$ denote conditions with $Q = 2Q_0$

Thus, $Q_0 = \frac{\kappa}{n_0} A_0 R_{h0}^{2/3} S_{00}$, where $\kappa = 1$, $A_0 = (0.9\,m)^2 = 0.81\,m^2$,

$P_0 = 2\sqrt{2}(0.9\,m) = 2.55\,m$ Thus, $R_{h0} = \frac{A_0}{P_0} = \frac{0.81\,m^2}{2.55\,m} = 0.318\,m$

Hence, $Q_0 = \frac{1}{n_0}(0.810)(0.318)^{2/3} S_{00}^{1/2}$

or

$$Q_0 = \frac{0.377\,S_{00}^{1/2}}{n_0} \quad (1)$$

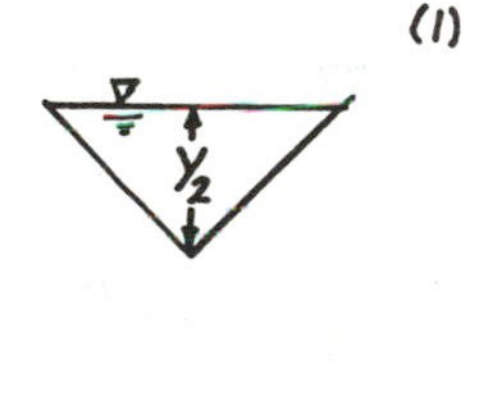

Also, $Q_2 = \frac{\kappa}{n_2} A_2 R_{h2}^{2/3} S_{02}$, where $A_2 = y_2^2$,

$P_2 = 2\sqrt{2}\,y_2$, or $R_{h2} = \frac{A_2}{P_2} = \frac{y_2^2}{2\sqrt{2}\,y_2} = 0.354\,y_2$

Hence, with $n_0 = n_2$ and $S_{00} = S_{02}$,

$$Q_2 = \frac{1}{n_0} y_2^2 (0.354\,y_2)^{2/3} S_{00}^{1/2}$$

or

$$Q_2 = \frac{0.500\,S_{00}^{1/2}}{n_0} y_2^{8/3} \quad (2)$$

From Eqs. (1) and (2) with $Q_2 = 2Q_0$ we obtain

$0.500\,y_2^{8/3} = 2(0.377)$ or $y_2 = 1.167\,m$

However,

$y_2 - 0.9\,m = \ell \sin 45°$ so that

$$\ell = \frac{y_2 - 0.9\,m}{\sin 45°} = \frac{1.167\,m - 0.9\,m}{\sin 45°} = \underline{\underline{0.378\,m}}$$

ℓ, y_2, 0.9m, 45°

10.9R

10.9R (Manning equation) Water flows in a rectangular channel of width b at a depth of $b/3$. Determine the diameter of a circular channel (in terms of b) that carries the same flowrate when it is half-full. Both channels have the same Manning coefficient, n, and slope.

(ANS: 0.889 b)

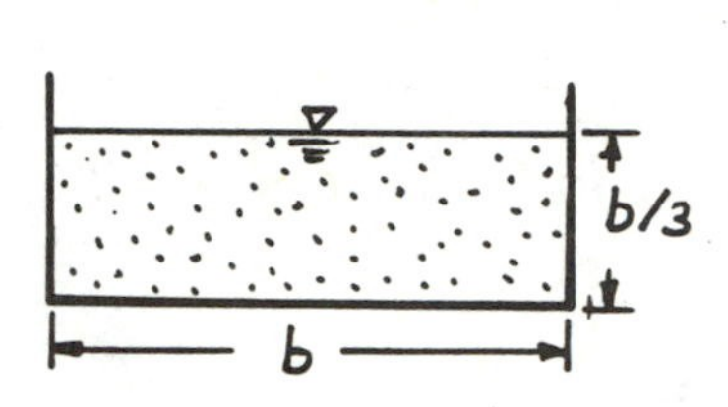

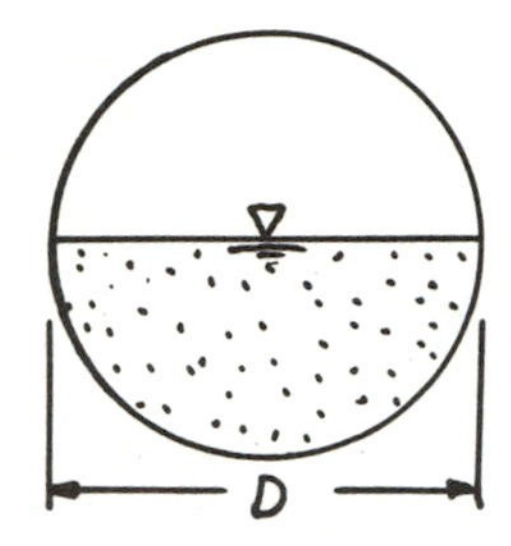

$$Q = \frac{\kappa}{n} A R_h^{2/3} S_o^{1/2} \qquad (1)$$

For the rectangle, $A = b\left(\frac{b}{3}\right)$ and $P = b + 2\left(\frac{b}{3}\right) = \frac{5}{3}b$

or $R_h = \frac{A}{P} = \frac{\frac{1}{3}b^2}{\frac{5}{3}b} = \frac{1}{5}b$

For the semi-circle, $A = \frac{\pi}{8}D^2$ and $P = \frac{\pi}{2}D$

or

$$R_h = \frac{1}{4}D$$

Thus, from Eq.(1) with equal Q, κ, n, and S_o for either case:

$$\left(\frac{b^2}{3}\right)\left(\frac{1}{5}b\right)^{2/3} = \left(\frac{\pi}{8}D^2\right)\left(\frac{1}{4}D\right)^{2/3}$$

or $\underline{\underline{D = 0.889\ b}}$

10.10R

10.10R (Manning equation) A weedy irrigation canal of trapezoidal cross section is to carry 20 m^3/s when built on a slope of 0.56 m/km. If the sides are at a 45° angle and the bottom is 8 m wide, determine the width of the waterline at the free surface.

(ANS: 12.0 m)

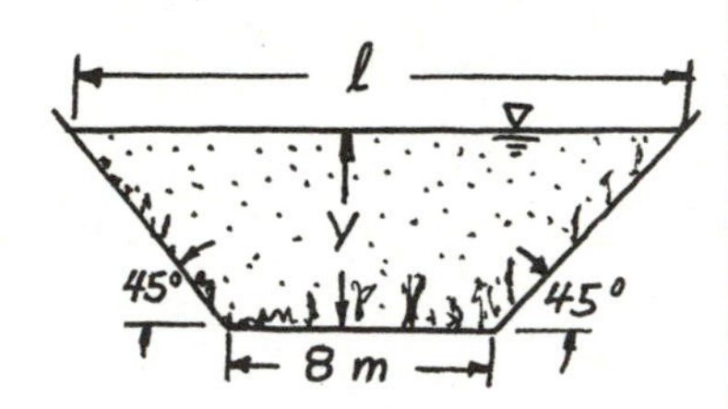

$$Q = \frac{\kappa}{n} A R_h^{2/3} S_o^{1/2}, \text{ where from Table 10.1, } n = 0.03 \quad (1)$$

Also, $\kappa = 1$ and $S_o = \frac{0.56\,m}{1000\,m} = 5.6\times10^{-4}$

$$A = \tfrac{1}{2}(\ell + 8m)y = \tfrac{1}{2}(2y + 8 + 8)y = (y+8)y$$

and

$$P = 8m + 2\frac{y}{\sin 45^\circ} = 8 + 2.83y$$

so that

$$R_h = \frac{A}{P} = \frac{(y+8)y}{(8+2.83y)}$$

Thus, Eq. (1) becomes

$$20 = \frac{1}{0.03}(y+8)y\left[\frac{(y+8)y}{(8+2.83y)}\right]^{2/3}(5.6\times10^{-4})^{1/2}$$

or

$$25.4 = \frac{(y^2+8y)^{5/3}}{(8+2.83y)^{2/3}}$$

or

$$y^2 + 8y - 6.96(8 + 2.83y)^{0.4} = 0 \equiv F(y) \quad (2)$$

Solve for $F(y) = 0$

From the graph, $F = 0$ when

$y \approx 1.98\ m$

Thus, $\ell = 2y + 8\ m$

or

$\ell = 2(1.98m) + 8m = \underline{\underline{12.0m}}$

10.11R

10.11R (Manning equation) Determine the maximum flowrate possible for the creek shown in Fig. P10.11R if it is not to overflow onto the floodplain. The creek bed drops an average of 5 ft/half mile of length. Determine the flowrate during a flood if the depth is 8 ft.

(ANS: 182 ft^3/s 1517 ft^3/s)

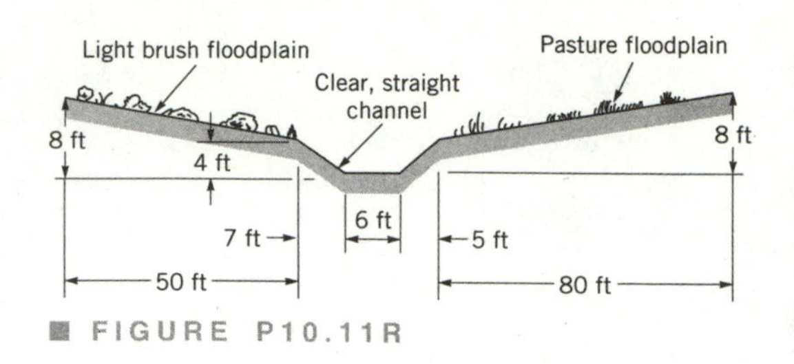

FIGURE P10.11R

(a) Maximum Q without overflowing:

$$Q = \frac{K}{n} A R_h^{2/3} S_o^{1/2}, \text{ where } K = 1.49, \; S_o = \frac{5\text{ ft}}{\left(\frac{5280}{2}\text{ft}\right)} = 0.00189, \qquad (1)$$

and from Table 10.1 $n = 0.030$

Also, $A = \frac{1}{2}(6\text{ft} + 18\text{ft})(4\text{ft}) = 48\text{ft}^2$

and $P = 6.40\text{ ft} + 6\text{ ft} + 8.06\text{ ft} = 20.5\text{ ft}$

so that $R_h = \frac{A}{P} = \frac{48\text{ ft}^2}{20.5\text{ ft}} = 2.34\text{ ft}$

(sketch: top width 18 ft, depth 4 ft, bottom 6 ft, sides ℓ_2 and ℓ_1)

$\ell_1 = (5^2 + 4^2)^{1/2} = 6.40\text{ft}$

$\ell_2 = (7^2 + 4^2)^{1/2} = 8.06\text{ ft}$

Thus, from Eq. (1)

$$Q = \frac{1.49}{0.030}(48)(2.34)^{2/3}(0.00189)^{1/2} = \underline{\underline{182 \frac{\text{ft}^3}{\text{s}}}}$$

(b) At a flood depth of 8 ft:

$Q = Q_1 + Q_2 + Q_3$, where

$Q_i = \frac{K}{n_i} A_i R_{h_i}^{2/3} S_o^{1/2}$ with

(sketch: 50 ft, 18 ft, 80 ft; regions (3), (1), (2); 8 ft; 4 ft; 6 ft; sides ℓ_3, ℓ_2, ℓ_1, ℓ_4)

$K = 1.49$, $S_o = 0.00189$, and from Table 10.1: $n_1 = 0.030$, $n_2 = 0.035$, $n_3 = 0.050$

From part (a), $\ell_1 = 6.40\text{ft}$ and $\ell_2 = 8.06\text{ ft}$

while $\ell_3 = (50^2 + 4^2)^{1/2} = 50.2\text{ ft}$ and $\ell_4 = (80^2 + 4^2)^{1/2} = 80.1\text{ ft}$

Thus, $A_1 = (4\text{ft})(18\text{ft}) + \frac{1}{2}(6\text{ft} + 18\text{ft})(4\text{ft}) = 120\text{ ft}^2$,

$A_2 = \frac{1}{2}(4\text{ft})(80\text{ft}) = 160\text{ ft}^2$, and $A_3 = \frac{1}{2}(4\text{ft})(50\text{ft}) = 100\text{ ft}^2$

and $P_1 = (6.40 + 6 + 8.06)\text{ ft} = 20.5\text{ ft}$, $P_2 = 80.1\text{ ft}$, and $P_3 = 50.2\text{ ft}$

Hence, since $R_{h_i} = \frac{A_i}{P_i}$ we obtain $A_i R_{h_i}^{2/3} = A_i^{5/3}/P_i^{2/3}$, or

$$Q = 1.49(0.00189)^{1/2}\left[\frac{(120)^{5/3}}{(20.5)^{2/3}(0.03)} + \frac{(160)^{5/3}}{(80.1)^{2/3}(0.035)} + \frac{(100)^{5/3}}{(50.2)^{2/3}(0.05)}\right]$$

$$= (842 + 470 + 205)\frac{\text{ft}^3}{\text{s}}$$

or

$$Q = \underline{\underline{1517 \frac{\text{ft}^3}{\text{s}}}}$$

10.12R (Best hydraulic cross section) Show that the triangular channel with the best hydraulic cross section (i.e., minimum area, A, for a given flowrate) is a right triangle as is shown in Fig. E10.8b.

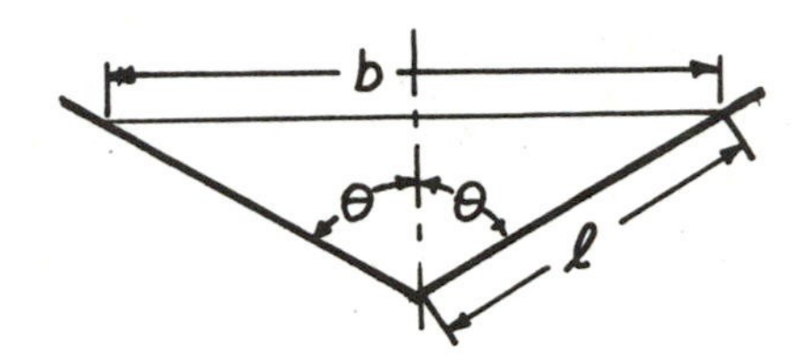

$Q = \frac{K}{n} A R_h^{2/3} S_o^{1/2}$, or with K, n, S_o constant

$$\frac{dQ}{d\theta} = \frac{K S_o^{1/2}}{n}\left[R_h^{2/3}\frac{dA}{d\theta} + A\left(\frac{2}{3}\right)R_h^{-1/3}\frac{dR_h}{d\theta}\right] \qquad (1)$$

Thus, for a given flowrate $\frac{dQ}{d\theta} = 0$ and for the minimum area $\frac{dA}{d\theta} = 0$ Eq. (1) gives

$$\frac{dR_h}{d\theta} = 0$$

Also, $R_h = \frac{A}{P}$, where $P = 2\ell$ and $A = \frac{1}{2}b(\ell\cos\theta)$

or

$$R_h = \frac{\frac{1}{2}b(\ell\cos\theta)}{2\ell} = \frac{1}{4}b\cos\theta \qquad (2)$$

However, since $\sin\theta = \frac{b}{2\ell}$ or $\ell = \frac{b}{2\sin\theta}$ it follows that

$$A = \frac{1}{2}b\left(\frac{b\cos\theta}{2\sin\theta}\right) \text{ or } b = 2\sqrt{A}\sqrt{\tan\theta}$$

Thus, Eq. (2) becomes:

$$R_h = \frac{\cos\theta}{4}\left(2\sqrt{A}\sqrt{\tan\theta}\right) = \frac{1}{2}\sqrt{A}\left(\sin\theta\cos\theta\right)^{1/2}$$

so that

$$\frac{dR_h}{d\theta} = \frac{1}{2}(\sin\theta\cos\theta)^{1/2}\frac{1}{2}A^{-1/2}\frac{dA}{d\theta} + \frac{1}{2}\sqrt{A}\left(\frac{1}{2}\right)(\sin\theta\cos\theta)^{-1/2}(\cos^2\theta - \sin^2\theta)$$

Thus, with $\frac{dA}{d\theta} = 0$ (i.e. minimum area), $\frac{dR_h}{d\theta} = 0$ when $\cos^2\theta - \sin^2\theta = 0$

or

$\underline{\underline{\theta = 45°}}$ (i.e. the best hydraulic cross-section occurs with a right triangle)

10.13R

10.13R (Hydraulic jump) At the bottom of a water ride in an amusement park, the water in the rectangular channel has a depth of 1.2 ft and a velocity of 15.6 ft/s. Determine the height of the "standing wave" (a hydraulic jump) that the boat passes through for its final "splash."

(ANS: 2.50 ft)

$V_1 = 15.6\,\frac{ft}{s}$ $\quad y_1 = 1.2\text{ ft}$ $\quad y_2$

The depth ratio for a hydraulic jump is

$$\frac{y_2}{y_1} = \frac{1}{2}\left[-1 + \sqrt{1 + 8\,Fr_1^2}\right], \text{ where } Fr_1 = \frac{V_1}{(g\,y_1)^{1/2}} = \frac{15.6\,\frac{ft}{s}}{\left[(32.2\,\frac{ft}{s^2})(1.2\,ft)\right]^{1/2}} = 2.51$$

Hence,

$$\frac{y_2}{y_1} = \frac{1}{2}\left[-1 + \sqrt{1 + 8(2.51)^2}\right] = 3.08 \text{ so that } y_2 = (3.08)(1.2\,ft) = 3.70\,ft$$

Thus, $y_2 - y_1 = (3.70 - 1.20)\,ft = \underline{\underline{2.50\,ft}}$

10.14R

10.14R (Hydraulic jump) Water flows in a rectangular channel with velocity $V = 6$ m/s. A gate at the end of the channel is suddenly closed so that a wave (a moving hydraulic jump) travels upstream with velocity $V_w = 2$ m/s as is indicated in Fig. P10.14R. Determine the depths ahead of and behind the wave. Note that this is an unsteady problem for a stationary observer. However, for an observer moving to the left with velocity V_w, the flow appears as a steady hydraulic jump.

(ANS: 0.652 m; 2.61 m)

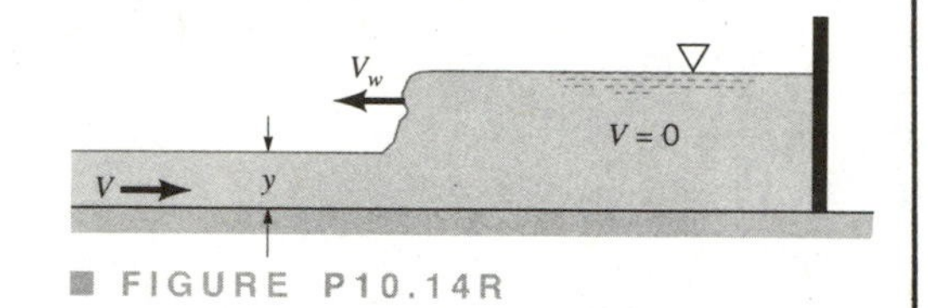

■ FIGURE P10.14R

For an observer moving to the left with speed $V_w = 2\,\frac{m}{s}$ the flow appears as shown below.

$V_1 = V + V_w = 8\,\frac{m}{s}$ $\quad y_2$ $\quad V_2 = V_w = 2\,\frac{m}{s}$

Thus, treat as a jump with

$V_1 = 8\,\frac{m}{s}$, $V_2 = 2\,\frac{m}{s}$

Since $A_1 V_1 = A_2 V_2$ or $\frac{y_2}{y_1} = \frac{V_1}{V_2} = \frac{8\,\frac{m}{s}}{2\,\frac{m}{s}} = 4$ it follows that

$$\frac{y_2}{y_1} = \frac{1}{2}\left[-1 + \sqrt{1 + 8\,Fr_1^2}\right] = 4 \qquad \text{Hence, } Fr_1 = 3.16$$

However, $Fr_1 = \frac{V_1}{(g\,y_1)^{1/2}}$ so that

$$y_1 = \frac{V_1^2}{g\,Fr_1^2} = \frac{(8\,\frac{m}{s})^2}{(9.81\,\frac{m}{s^2})(3.16)^2} = \underline{\underline{0.652\,m}}$$

and

$$y_2 = 4\,y_1 = 4(0.652\,m) = \underline{\underline{2.61\,m}}$$

10.15R

10.15R (Sharp-crested weir) Determine the head, H, required to allow a flowrate of 600 m^3/hr over a sharp-crested triangular weir with $\theta = 60°$.

(ANS: 0.536 m)

$Q = C_{wt} \frac{8}{15} \tan\frac{\theta}{2} \sqrt{2g}\, H^{5/2}$, where $Q = 600 \frac{m^3}{hr}\left(\frac{1 hr}{3600 s}\right) = 0.167 \frac{m^3}{s}$ and $\theta = 60°$

Thus, $0.167 \frac{m^3}{s} = C_{wt} \frac{8}{15}(\tan 30°)\left[2(9.81\frac{m}{s^2})\right]^{1/2} H^{5/2}$

or
$$0.122 = C_{wt} H^{5/2} \qquad (1)$$

The value of C_{wt} is obtained from Fig. 10.25 as a function of H.

$\theta = 60°$, C_{wt}, 0, H, ft, 1 — Fig. 10.25 (2)

Trial and error solution of Eqs. (1), (2):

Assume $C_{wt} = 0.60$, or
$0.122 = 0.60 H^{5/2}$ so that $H = 0.529$ m
or $H = (0.529 m)(3.28 \frac{ft}{m}) = 1.73$ ft

From Fig. 10.25 this gives $C_{wt} \approx 0.58$ (extrapolate for $H > 1$ ft)

Assume $C_{wt} = 0.58$, or
$0.122 = 0.58 H^{5/2}$ so that $H = (0.536 m)(3.28 \frac{ft}{m}) = 1.76$ ft
which checks with $C_{wt} = 0.58$, the assumed value.

Thus,

$H = \underline{\underline{0.536\ m}}$

10.16R

10.16R (Broad-crested weir) The top of a broad-crested weir block is at an elevation of 724.5 ft, which is 4 ft above the channel bottom. If the weir is 20-ft wide and the flowrate is 400 cfs, determine the elevation of the reservoir upstream of the weir.

(ANS: 730.86 ft)

(1)

Q = 400 cfs

H

Z = 724.5 ft

$P_w = 4$ ft

b = 20 ft = width

$$Q = \frac{0.65}{\left(1+\frac{H}{P_w}\right)^{1/2}}\, b\sqrt{g}\left(\frac{2}{3}\right)^{3/2} H^{3/2}$$, or with $P_w = 4$ ft, $b = 20$ ft, and $Q = 400$ cfs,

$$400\,\frac{ft^3}{s} = \frac{0.65}{\left(1+\frac{H}{4}\right)^{1/2}}\,(20\,ft)\sqrt{32.2\,\frac{ft}{s^2}}\left(\frac{2}{3}\right)^{3/2} H^{3/2}$$, where $H \sim ft$

or

$$9.96\left(1+\frac{H}{4}\right)^{1/2} = H^{3/2}$$

Thus,

$$H^3 - 24.8H - 99.2 = 0$$

The roots of this cubic equation are $H = 6.36$ and two complex roots. Thus, $H = 6.36$ ft so that

$$z_1 = 724.5\,ft + 6.36\,ft = \underline{\underline{730.86\,ft}}$$

10.17R

10.17R (Underflow gate) Water flows under a sluice gate in a 60-ft-wide finished concrete channel as is shown in Fig. P10.17R. Determine the flowrate. If the slope of the channel is 2.5 ft/100 ft, will the water depth increase or decrease downstream of the gate? Assume $C_c = y_2/a = 0.65$. Explain.

(ANS: 1670 ft^3/s; decrease)

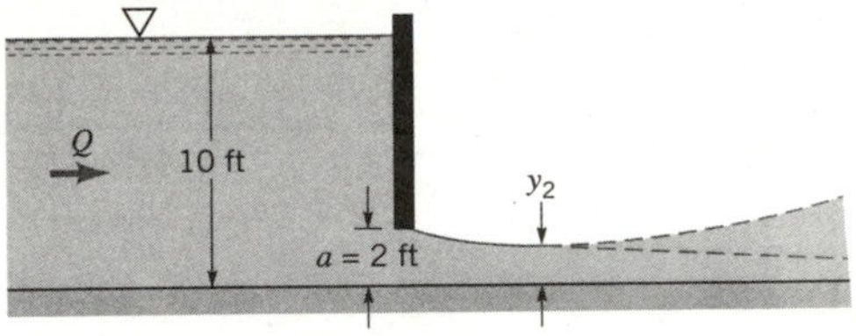

■ FIGURE P10.17R

$Q = bq = b\,C_d\,a\sqrt{2gy_1}$, where $b = 60\,ft$, $a = 2\,ft$, and from Fig. 10.29

since $\frac{y_1}{a} = \frac{10\,ft}{2\,ft} = 5$, it follows that $C_d = 0.55$

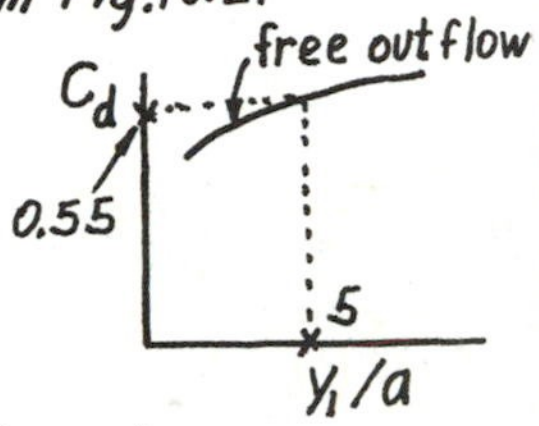

Hence,

$$Q = (60\,ft)(0.55)(2\,ft)\left[2\left(32.2\,\frac{ft}{s^2}\right)(10\,ft)\right]^{1/2} = \underline{\underline{1670\,\frac{ft^3}{s}}}$$

Determine the slope needed to maintain uniform flow downstream of the gate:

$Q = \frac{\kappa}{n} A R_h^{2/3} S_o^{1/2}$, where $\kappa = 1.49$ and from Table 10.1 $n = 0.012$ (1)

Also, $y_2 = C_c\,a = 0.65\,(2\,ft) = 1.3\,ft$

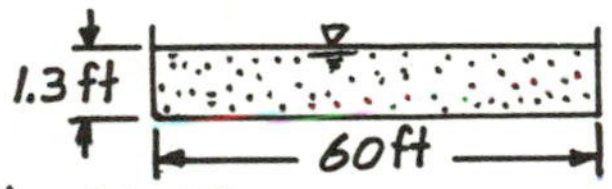

so that

$A = (1.3\,ft)(60\,ft) = 78\,ft^2$, $P = (60 + 2(1.3))\,ft = 62.6\,ft$

and

$R_h = \frac{A}{P} = \frac{78\,ft^2}{62.6\,ft} = 1.245\,ft$

Thus, from Eq. (1):

$$1670 = \frac{1.49}{0.012}(78)(1.245)^{2/3} S_o^{1/2}, \text{ or } S_o = 0.0222$$

Hence, the required slope for uniform flow is $S_o = 0.0222$, but the actual slope is $S_o = \frac{2.5\,ft}{100\,ft} = 0.0250$, more than required. The fluid will speed up and the <u>depth decrease</u>.

11 Compressible Flow

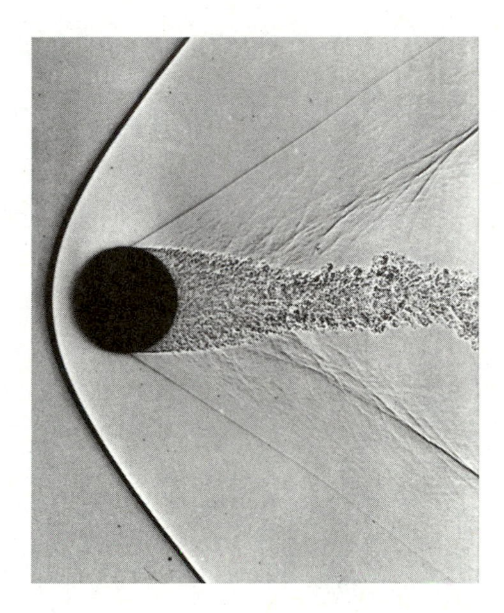

Flow past a sphere at Mach 1.53: An object moving through a fluid at supersonic speed (Mach number greater than one) creates a shock wave (a discontinuity in flow conditions shown by the dark curved line), which is heard as a sonic boom as the object passes overhead. The turbulent wake is also shown (shadowgraph technique used in air). (Photography courtesy of A. C. Charters.)

11.1R

11.1R (Speed of sound) Determine the speed of sound in air for a hot summer day when the temperature is 100 °F; for a cold winter day when the temperature is −20 °F.

(ANS: 1160 ft/s; 1028 ft/s)

For a perfect gas $c = \sqrt{kRT}$, where for air $k = 1.4$ and $R = 1716 \frac{ft \cdot lb}{slug \cdot {}^\circ R}$ (see Table 1.7).

Hence,

$$c_{100^\circ F} = \sqrt{1.4 \left(1716 \frac{ft \cdot lb}{slug \cdot {}^\circ R}\right)(460 + 100)^\circ R} = 1160 \sqrt{\frac{ft \cdot lb}{slug}} = 1160 \sqrt{\frac{ft (slug \cdot ft/s^2)}{slug}}$$

or

$$c_{100^\circ F} = \underline{\underline{1160 \frac{ft}{s}}} = 791 \; mph$$

Similarly,

$$c_{-20^\circ F} = \sqrt{1.4 \left(1716 \frac{ft \cdot lb}{slug \cdot {}^\circ R}\right)(460 - 20)^\circ R} = \underline{\underline{1028 \frac{ft}{s}}} = 701 \; mph$$

11.2R

11.2R (Speed of sound) Compare values of the speed of sound in ft/s in the following liquids at 68 °F: **(a)** ethyl alcohol, **(b)** glycerin, **(c)** mercury.

(ANS: 3810 ft/s; 6220 ft/s; 4760 ft/s)

For a liquid $c = \sqrt{\frac{E_v}{\rho}}$ where the values of E_v (bulk modulus) and ρ (density) are given in Table 1.5.

Thus, for ethyl alcohol

$$c_{al} = \left[\frac{\left(1.54 \times 10^5 \frac{lb}{in.^2}\right)\left(144 \frac{in.^2}{ft^2}\right)}{1.53 \frac{slugs}{ft^3}} \right]^{\frac{1}{2}} = 3810 \left[\frac{ft \cdot lb}{slug} \right]^{\frac{1}{2}} = 3810 \left[\frac{ft \cdot (slug \cdot ft/s^2)}{slug} \right]^{\frac{1}{2}}$$

or

$$c_{al} = \underline{\underline{3810 \frac{ft}{s}}} = 2600 \; mph$$

For glycerin

$$c_{gl} = \left[\frac{\left(6.56 \times 10^5 \frac{lb}{in.^2}\right)\left(144 \frac{in.^2}{ft^2}\right)}{2.44 \frac{slugs}{ft^3}} \right]^{\frac{1}{2}} = \underline{\underline{6\,220 \frac{ft}{s}}} = 4240 \; mph$$

and for mercury

$$c_m = \left[\frac{\left(4.14 \times 10^6 \frac{lb}{in.^2}\right)\left(144 \frac{in.^2}{ft^2}\right)}{26.3 \frac{slugs}{ft^3}} \right]^{\frac{1}{2}} = \underline{\underline{4760 \frac{ft}{s}}} = 3250 \; mph$$

11.3R

11.3R (Sound waves) A stationary point source emits weak pressure pulses in a flow that moves uniformly from left to right with a Mach number of 0.5. Sketch the instantaneous outline at time = 10s of pressure waves emitted earlier at time = 5s and time = 8s. Assume the speed of sound is 1000 ft/s.

Since $Ma = \frac{V}{c}$ it follows that the fluid speed is

$V = c\,Ma = 1000\,\frac{ft}{s}(0.5) = 500\,\frac{ft}{s}$.

Assume that the sound source is located at $x=y=0$.

At time $t=10s$, the pulse that was emitted at $t=8s$ has traveled a distance $R_8 = c\,\Delta t = 1000\,\frac{ft}{s}(10s-8s) = 2000\,ft$ relative to the moving fluid that has moved a distance $x_8 = V\Delta t = 500\,\frac{ft}{s}(10s-8s) = 1000\,ft$. Similarly, for the pulse that was emitted at $t=5s$, $R_5 = 1000\,\frac{ft}{s}(10s-5s) = 5000\,ft$ and $x_5 = 500\,\frac{ft}{s}(10s-5s) = 2500\,ft$.

These pulses at $t=10s$ are shown below. They are circles.

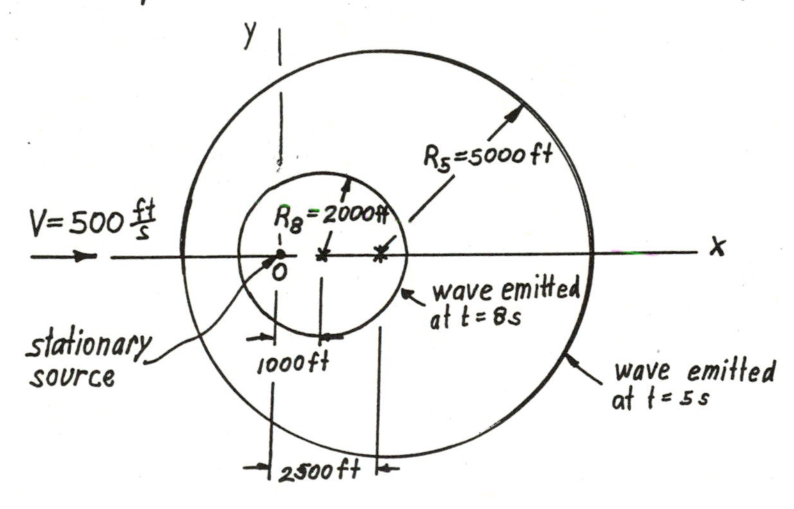

11.4R

11.4R (Mach number) An airplane moves forward in air with a speed of 500 mph at an altitude of 40,000 ft. Determine the Mach number involved if the air is considered as U.S. standard atmosphere (see Table C.1).

(ANS: 0.757)

$Ma = \frac{V}{c}$, where $c = \sqrt{kRT}$

From Table C.1, the temperature at 40,000 ft is $-69.7°F$.
Thus,

$$c = \left[1.4\left(1716\,\frac{ft \cdot lb}{slug \cdot °R}\right)(460 - 69.7)°R\right]^{\frac{1}{2}} = 968\,\frac{ft}{s}$$

so that

$$Ma = \frac{\left(500\,\frac{mi}{hr}\right)\left(\frac{1\,hr}{3600\,s}\right)\left(\frac{5280\,ft}{1\,mi}\right)}{968\,\frac{ft}{s}} = \underline{\underline{0.757}}$$

11.5R

11.5R (Isentropic flow) At section (1) in the isentropic flow of carbon dioxide, $p_1 = 40$ kPa(abs), $T_1 = 60$ °C, and $V_1 = 350$ m/s. Determine the flow velocity, V_2, in m/s, at another section, section (2), where the Mach number is 2.0. Also calculate the section area ratio, A_2/A_1.

(ANS: 500 m/s; 1.71)

$p_1 = 40$ kPa(abs), $V_1 = 350$ m/s, $T_1 = 60°C$, CO_2, V_2, $Ma_2 = 2$, (1), (2)

$$V_2 = Ma_2\, c_2 = 2\, c_2 = 2\sqrt{kRT_2} \qquad (1)$$

where for CO_2, $k = 1.30$ and $R = 188.9 \frac{N \cdot m}{kg \cdot K}$ (see Table 1.8)

Also, from Eq. 11.56

$$\frac{T_2}{T_{0,2}} = \frac{1}{1+\left(\frac{k-1}{2}\right)Ma_2^2} \text{ and } \frac{T_1}{T_{0,1}} = \frac{1}{1+\left(\frac{k-1}{2}\right)Ma_1^2}$$, where for isentropic flow, $T_{0,1} = T_{0,2}$. Thus, by dividing these equations we obtain

$$\frac{T_2}{T_1} = \frac{1+\left(\frac{k-1}{2}\right)Ma_1^2}{1+\left(\frac{k-1}{2}\right)Ma_2^2} \qquad (2)$$

However,

$$Ma_1 = \frac{V_1}{\sqrt{kRT_1}} = \frac{350 \frac{m}{s}}{\sqrt{188.9 \frac{N \cdot m}{kg \cdot K}(1.3)(273+60)K}} = 1.22$$

Thus, from Eq. (2),

$$T_2 = (273+60)K \frac{\left[1+\left(\frac{1.3-1}{2}\right)(1.22)^2\right]}{\left[1+\left(\frac{1.3-1}{2}\right)(2)^2\right]} = 255\,K, \text{ so from Eq. (1)}$$

$$V_2 = 2\left[1.3\left(188.9 \frac{N \cdot m}{kg \cdot K}\right)(255)\right]^{1/2} = \underline{\underline{500 \frac{m}{s}}}$$

Also, (see Eq. 11.71)

$$\frac{A_2}{A_1} = \frac{\left(\frac{A_2}{A^*}\right)}{\left(\frac{A_1}{A^*}\right)} = \frac{\frac{1}{Ma_2}\left[\frac{1+\left(\frac{k-1}{2}\right)Ma_2^2}{1+\left(\frac{k-1}{2}\right)}\right]^{\frac{k+1}{2(k-1)}}}{\frac{1}{Ma_1}\left[\frac{1+\left(\frac{k-1}{2}\right)Ma_1^2}{1+\left(\frac{k-1}{2}\right)}\right]^{\frac{k+1}{2(k-1)}}} = \frac{Ma_1}{Ma_2}\left[\frac{1+\left(\frac{k-1}{2}\right)Ma_2^2}{1+\left(\frac{k-1}{2}\right)Ma_1^2}\right]^{\frac{k+1}{2(k-1)}}$$

so that with

$\frac{k-1}{2} = \frac{1.3-1}{2} = 0.15$, $\frac{k+1}{2(k-1)} = \frac{1.3+1}{2(1.3-1)} = 3.83$, $Ma_1 = 1.22$, and $Ma_2 = 2$

we obtain

$$\frac{A_2}{A_1} = \frac{1.22}{2}\left[\frac{1+0.15(2)^2}{1+0.15(1.22)^2}\right]^{3.83} = \underline{\underline{1.71}}$$

11.6R

11.6R (Isentropic flow) An ideal gas in a large storage tank at 100 psia and 60 °F flows isentropically through a converging duct to the atmosphere. The throat area of the duct is 0.1 ft^2. Determine the pressure, temperature, velocity, and mass flowrate of the gas at the duct throat if the gas is **(a)** air, **(b)** carbon dioxide, **(c)** helium.

(ANS: **(a)** 52.8 psia; 433 °R; 1020 ft/s; 1.04 slugs/s; **(b)** 54.6 psia; 452 °R; 815 ft/s; 1.25 slugs/s; **(c)** 48.8 psia; 391 °R; 2840 ft/s; 0.411 slugs/s)

$V_0 = 0$
$p_0 = 100\,\text{psia}$
$T_0 = 60\,°F$
$A_{th} = 0.1\,ft^2$
V_{th}
(th) p_{atm}

If the nozzle is choked, then the fluid speed at the exit is equal to the speed of sound there and the pressure is the critical pressure, p^*, where from Eq. 11.61

$$\frac{p^*}{p_0} = \left(\frac{2}{k+1}\right)^{k/(k-1)}$$

If $p^* < p_{atm}$, then the nozzle is not choked and $p_{th} = p_{atm}$. If $p^* > p_{atm}$, then the nozzle is choked and $p_{th} = p^*$. As shown below, the nozzle is choked in each case considered.

a) For air $k = 1.4$ so that

$$p^* = (100\,\text{psia})\left(\frac{2}{1.4+1}\right)^{1.4/(1.4-1)} = 52.8\,\text{psia} > p_{atm} = 14.7\,\text{psia}$$

Thus, $p_{th} = p^* = \underline{\underline{52.8\,\text{psia}}}$

From Eq. 11.63, since $Ma_{th} = 1$ we have $T_{th} = T^* = T_0\left(\frac{2}{k+1}\right)$

or

$$T_{th} = (460+60)\left(\frac{2}{1.4+1}\right) = \underline{\underline{433\,°R}} \quad \text{so that}$$

$$V_{th} = Ma_{th}\, c_{th} = \sqrt{kRT_{th}} = \left[1.4\left(1716\,\frac{ft\cdot lb}{slug\cdot °R}\right)(433\,°R)\right]^{\frac{1}{2}} = \underline{\underline{1020\,\frac{ft}{s}}}$$

Finally,

$$\dot{m} = \rho A V = \left(\frac{p_{th}}{R\,T_{th}}\right) A_{th} V_{th} = \frac{(52.8\,\frac{lb}{in.^2})(144\,\frac{in.^2}{ft^2})(0.1\,ft^2)(1020\,\frac{ft}{s})}{(1716\,\frac{ft\cdot lb}{slug\cdot °R})(433\,°R)}$$

or

$$\dot{m} = \underline{\underline{1.04\,\frac{slugs}{s}}}$$

(continued)

b) Similarly for carbon dioxide $k = 1.3$ so that

$$p^* = (100\,psia)\left(\frac{2}{1.3+1}\right)^{1.3/(1.3-1)} = \underline{\underline{54.6\,psia}} > p_{atm} \text{ so } p_{th} = p^*$$

Also

$$T_{th} = T^* = T_0\left(\frac{2}{k+1}\right) = (460+60)\left(\frac{2}{1.3+1}\right) = \underline{\underline{452\,°R}} \text{ so that}$$

$$V_{th} = \sqrt{kRT_{th}} = \left[1.3\left(1130\,\frac{ft\cdot lb}{slug\cdot °R}\right)(452\,°R)\right]^{1/2} = \underline{\underline{815\,\frac{ft}{s}}}$$

and

$$\dot{m} = \left(\frac{p_{th}}{RT_{th}}\right)A_{th}V_{th} = \frac{\left(54.6\,\frac{lb}{in.^2}\right)\left(144\,\frac{in.^2}{ft^2}\right)(0.1\,ft^2)\left(815\,\frac{ft}{s}\right)}{\left(1130\,\frac{ft\cdot lb}{slug\cdot °R}\right)(452\,°R)} = \underline{\underline{1.25\,\frac{slugs}{s}}}$$

c) Finally, for helium $k = 1.66$ so that

$$p^* = (100\,psia)\left(\frac{2}{1.66+1}\right)^{1.66/(1.66-1)} = \underline{\underline{48.8\,psia}} > p_{atm} \text{ so that } p_{th} = p^*$$

Also,

$$T_{th} = T^* = T_0\left(\frac{2}{k+1}\right) = (460+60)\left(\frac{2}{1.66+1}\right) = \underline{\underline{391\,°R}} \text{ so that}$$

$$V_{th} = \sqrt{kRT_{th}} = \left[1.66\left(12{,}420\,\frac{ft\cdot lb}{slug\cdot °R}\right)(391\,°R)\right]^{1/2} = \underline{\underline{2840\,\frac{ft}{s}}}$$

and

$$\dot{m} = \left(\frac{p_{th}}{RT_{th}}\right)A_{th}V_{th} = \frac{\left(48.8\,\frac{lb}{in.^2}\right)\left(144\,\frac{in.^2}{ft^2}\right)(0.1\,ft^2)\left(2840\,\frac{ft}{s}\right)}{\left(12{,}420\,\frac{ft\cdot lb}{slug\cdot °R}\right)(391\,°R)} = \underline{\underline{0.411\,\frac{slugs}{s}}}$$

11.7R

11.7R (Fanno flow) A long, smooth wall pipe ($f = 0.01$) is to deliver 8000 ft^3/min of air at 60 °F and 14.7 psia. The inside diameter of the pipe is 0.5 ft and the length of the pipe is 100 ft. Determine the static temperature and pressure required at the pipe entrance if the flow through the pipe is adiabatic.

(ANS: 539 °R; 23.4 psia)

$\ell = 100$ ft

(1) (2)

$f = 0.01$ $D = 0.5$ ft $Q_2 = 8000 \frac{ft^3}{min}$

$p_2 = 14.7$ psia

$T_2 = 60°F$

At the exit

$$V_2 = \frac{Q_2}{A_2} = \frac{4Q_2}{\pi D^2} = \frac{4(8000\frac{ft^3}{min})(\frac{1 min}{60 s})}{\pi(0.5 ft)^2} = 679 \frac{ft}{s}$$

so that

$$Ma_2 = \frac{V_2}{c_2} = \frac{V_2}{\sqrt{kRT_2}} = \frac{679\frac{ft}{s}}{\left[1.4(1716\frac{ft\cdot lb}{slug\cdot°R})(460+60)°R\right]^{1/2}} = 0.608$$

Thus, from Fig. D.2 (Fanno flow) with $Ma_2 = 0.608$ we obtain

$\frac{f(\ell^* - \ell_2)}{D} = 0.45$ so that from

$$\frac{f(\ell_2 - \ell_1)}{D} = \frac{(0.01)(100 ft)}{0.5 ft} = 2.00 = \frac{f(\ell^* - \ell_1)}{D} - \frac{f(\ell^* - \ell_2)}{D}$$

we obtain

$$\frac{f(\ell^* - \ell_1)}{D} = 2.00 + 0.45 = 2.45$$

Hence, from Fig. D.2 we find $Ma_1 = 0.39$, $\frac{T_1}{T^*} = 1.16$, and $\frac{p_1}{p^*} = 2.75$

Thus, with

$T_1 = T_2\left(\frac{T^*}{T_2}\right)\left(\frac{T_1}{T^*}\right)$ and $p_1 = p_2\left(\frac{p^*}{p_2}\right)\left(\frac{p_1}{p^*}\right)$ we can obtain T_1 and p_1.

From Fig. D.2 with $Ma_2 = 0.608$ we have

$\frac{T_2}{T^*} = 1.12$ and $\frac{p_2}{p^*} = 1.73$

Thus,

$$T_1 = (460+60)°R \frac{(1.16)}{(1.12)} = \underline{\underline{539°R}}$$

and

$$p_1 = 14.7 \text{ psia} \frac{(2.75)}{(1.73)} = \underline{\underline{23.4 \text{ psia}}}$$

11.8R

11.8R (Rayleigh flow) Air enters a constant-area duct that may be considered frictionless with $T_1 = 300$ K and $V_1 = 300$ m/s. Determine the amount of heat transfer in kJ/kg required to choke the Rayleigh flow involved.
(ANS: 5020 J/kg)

$V_1 = 300\frac{m}{s}$ $q_{net\,in}$ $Ma_2 = 1$
(1) $f=0$ (2)
$T_1 = 300K$

For Rayleigh flow the energy equation (Eq. 5.69) is

$$\dot{m}\left[\check{h}_2 - \check{h}_1 + \frac{1}{2}(V_2^2 - V_1^2)\right] = \dot{Q}_{net\,in},$$

or

$$q_{net\,in} = \frac{\dot{Q}_{net\,in}}{\dot{m}} = h_{o2} - h_{o1} \text{ where } h_o = \check{h} + \frac{1}{2}V^2 = c_p T_o \text{ is}$$

the stagnation enthalpy. Thus,

$$q_{net\,in} = c_p(T_{o2} - T_{o1}) \qquad (1)$$

At section (1), $Ma_1 = \frac{V_1}{c_1} = \frac{V_1}{\sqrt{kRT_1}} = \frac{300\frac{m}{s}}{\left[1.4(286.9\frac{N\cdot m}{kg\cdot K})(300K)\right]^{1/2}} = 0.864$

Thus, from Eq. (11.56),

$$\frac{T_1}{T_{o1}} = \frac{1}{1 + \frac{(k-1)}{2}Ma_1^2}, \text{ or } T_{o1} = (300K)\left[1 + \frac{(1.4-1)}{2}(0.864)^2\right] = 345\,K$$

Since the flow is choked (i.e. $Ma_2 = 1$), $T_{o2} = T_{oa}$ where from Eq. (11.131),

$$\frac{T_{o1}}{T_{oa}} = \frac{2(k+1)\left(1 + \frac{k-1}{2}Ma_1^2\right)Ma_1^2}{(1 + k\,Ma_1^2)^2}, \text{ or}$$

$$T_{oa} = (345K)\frac{(1 + 1.4(0.864)^2)^2}{2(1.4+1)\left[1 + \frac{(1.4-1)}{2}(0.864)^2\right](0.864)^2} = 350K$$

Thus,

$T_{o2} = T_{oa} = 350\,K$ so that with $c_p = \frac{Rk}{k-1} = \frac{1.4(286.9\frac{N\cdot m}{kg\cdot K})}{(1.4-1)} = 1004\frac{J}{kg\cdot K}$

Eq. (1) gives

$$q_{net\,in} = (1004\tfrac{J}{kg\cdot K})(350K - 345K) = \underline{\underline{5020\tfrac{J}{kg}}}$$

11.9R

11.9R (Normal shock waves) Standard atmospheric air enters subsonically and accelerates isentropically to supersonic flow in a duct. If the ratio of duct exit to throat cross-section areas is 3.0, determine the ratio of back pressure to inlet stagnation pressure that will result in a standing normal shock at the duct exit. Determine also the stagnation pressure loss across the normal shock in kPa.

(ANS: 0.375; 56.1 kPa)

$p_0 = 101\ kPa$, (0), $V_0 = 0$; throat; normal shock; $Ma<1$; $Ma=1$; $Ma>1$; (*); x y; p_b (b); $A_x = 3A^*$

For isentropic flow with $\frac{A_x}{A^*} = 3.0$ we obtain Ma_x from Fig. D.1 or Eq. (11.71):

$Ma_x = 2.64$

Then from Eq. (11.59) we obtain

$$\frac{p_x}{p_{ox}} = \left[\frac{1}{1+\left(\frac{k-1}{2}\right)Ma_x^2}\right]^{k/(k-1)} = \left[\frac{1}{1+\left(\frac{1.4-1}{2}\right)(2.64)^2}\right]^{1.4/(1.4-1)} = 0.0471$$

while

$$\frac{p_b}{p_{o1}} = \frac{p_y}{p_{ox}} = \frac{p_y}{p_x}\frac{p_x}{p_{ox}}$$, where from Eq. (11.150) (or Fig. D.4)

$$\frac{p_y}{p_x} = \frac{2k}{(k+1)}Ma_x^2 - \frac{k-1}{k+1} = \frac{2(1.4)}{(1.4+1)}(2.64)^2 - \frac{(1.4-1)}{(1.4+1)} = 7.96$$

Thus, since $p_0 = p_{ox} = p_{o1}$

$$\frac{p_b}{p_0} = (7.96)(0.0471) = \underline{\underline{0.375}}, \quad \text{or } p_b = 101\,kP(0.375) = 37.9\,kPa$$

Finally, from Eq. (11.156) (or Fig. D.4)

$$\frac{p_{oy}}{p_{ox}} = \frac{\left(\frac{k+1}{2}Ma_x^2\right)^{k/(k-1)}\left(1+\frac{k-1}{2}Ma_x^2\right)^{k/(1-k)}}{\left(\frac{2k}{k+1}Ma_x^2 - \frac{k-1}{k+1}\right)^{1/(k-1)}}$$

$$= \frac{\left[\left(\frac{1.4+2}{2}\right)(2.64)^2\right]^{[1.4/0.4]}\left[1+\frac{0.4}{2}(2.64)^2\right]^{-\frac{1.4}{0.4}}}{\left[\frac{2.8}{2.4}(2.64)^2 - \frac{0.4}{2.4}\right]^{1/0.4}} = 0.445$$

Thus,

$$p_{ox} - p_{oy} = p_{ox}\left[1 - \frac{p_{oy}}{p_{ox}}\right] = 101\,kPa\,(abs)\left[1-0.445\right] = \underline{\underline{56.1\,kPa}}$$

12
Turbomachines

Laser velocimeter measurements of the flow field in the rotor row of a low-speed research turbine. (Photograph courtesy of Dr. D. C. Wisler, Director, Aerodynamic Research Laboratory of GE Aircraft Engines.)

12.1R

12.1R (Angular momentum) Water is supplied to a dishwasher through the manifold shown in Fig. P12.1R. Determine the rotational speed of the manifold if bearing friction and air resistance are neglected. The total flowrate of 2.3 gpm is divided evenly among the six outlets, each of which produces a 5/16-in.-diameter stream.

(ANS: 0.378 rev/s)

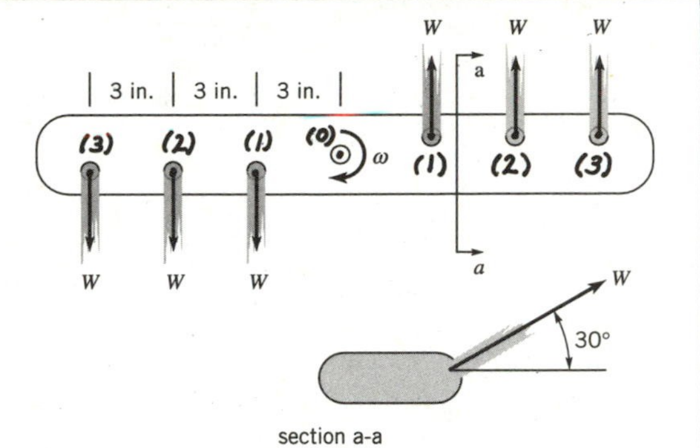

■ FIGURE P12.1R

With points (0), (1), (2), and (3) located as in the diagram above,

$T = \text{"}\dot{m}(r_{out}V_{\theta out} - r_{in}V_{\theta in})\text{"}$ where $V_{\theta in} = 0$. Thus,

$T = 2\dot{m}_1 r_1 V_{\theta 1} + 2\dot{m}_2 r_2 V_{\theta 2} + 2\dot{m}_3 r_3 V_{\theta 3} = 0$ since there is no friction.

But $\dot{m}_1 = \dot{m}_2 = \dot{m}_3$ so that the above becomes

$$r_1 V_{\theta 1} + r_2 V_{\theta 2} + r_3 V_{\theta 3} = 0 \qquad (1)$$

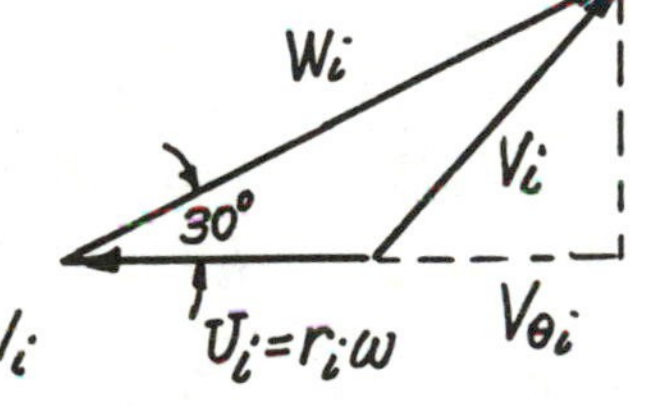

But $U_i + V_{\theta i} = W_i \cos 30°$, $i = 1,2,3$ (2)

where

$$Q = 6A_iW_i = 6\left[\frac{\pi}{4}\left(\frac{(\frac{5}{16})}{12}\ \text{ft}\right)^2\right]W_i = 0.00320\,W_i$$

with

$$Q = \left(2.3\frac{\text{gal}}{\text{min}}\right)\left(\frac{1\text{min}}{60\text{s}}\right)\left(231\frac{\text{in.}^3}{\text{gal}}\right)\left(\frac{1\text{ft}^3}{1728\text{in.}^3}\right) = 0.00512\frac{\text{ft}^3}{\text{s}}$$

Thus, $W_i = 1.60\ \frac{\text{ft}}{\text{s}}$ so that from Eq. (2) $V_{\theta i} = W_i \cos 30° - U_i$

or $V_{\theta i} = (1.60\ \frac{\text{ft}}{\text{s}})\cos 30° - r_i\omega$. With $r_1 = \frac{3}{12}$ ft, $r_2 = \frac{6}{12}$ ft, and $r_3 = \frac{9}{12}$ ft

Eq. (1) becomes

$$\frac{3}{12}\left(1.386 - \frac{3}{12}\omega\right) + \frac{6}{12}\left(1.386 - \frac{6}{12}\omega\right) + \frac{9}{12}\left(1.386 - \frac{9}{12}\omega\right) = 0$$

or $24.9 = 10.5\,\omega$

Thus, $\omega = 2.37\frac{\text{rad}}{\text{s}} \times \frac{1\,\text{rev}}{2\pi\,\text{rad}} = \underline{\underline{0.378\ \frac{\text{rev}}{\text{s}}}}$

12.2R

12.2R (Velocity triangles) An axial-flow turbomachine rotor involves the upstream (1) and downstream (2) velocity triangles shown in Fig. P12.2R. Is this turbomachine a turbine or a fan? Sketch an appropriate blade section and determine the energy transferred per unit mass of fluid.
(ANS: turbine; $-36.9\ ft^2/s^2$)

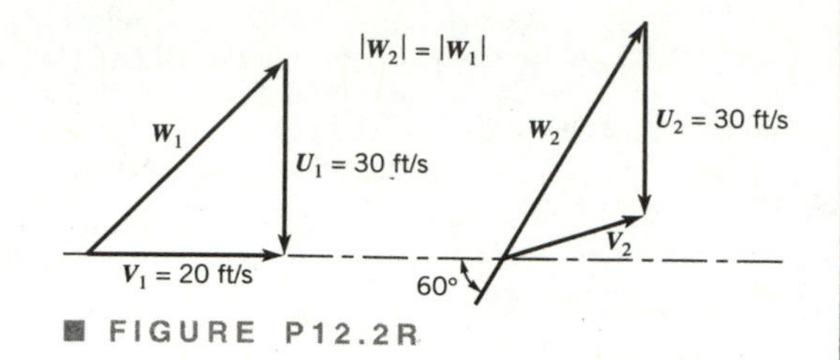

■ FIGURE P12.2R

From the given inlet conditions $W_1 = \sqrt{(30\frac{ft}{s})^2 + (20\frac{ft}{s})^2} = 36.06\frac{ft}{s}$

Thus, with $W_2 = W_1 = 36.06\frac{ft}{s}$ the outlet velocity triangle is as shown below,

or

$V_{\theta 2} = (36.06\frac{ft}{s})\sin 60° - 30\frac{ft}{s} = 1.229\frac{ft}{s}$

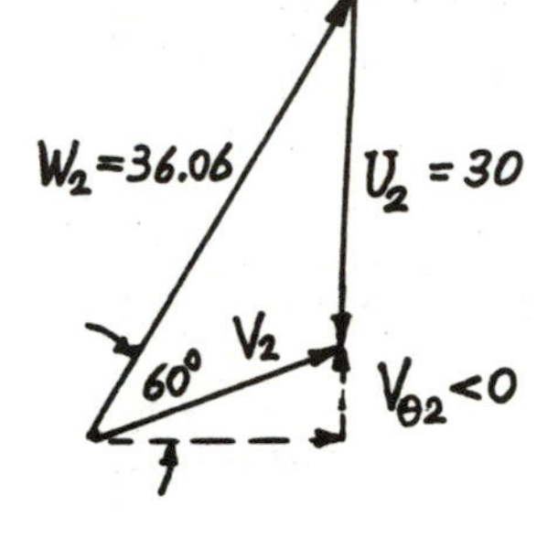

Thus, since

$T = \dot{m}(r_2 V_{\theta 2} - r_1 V_{\theta 1})$ with $V_{\theta 1} = 0$ and $V_{\theta 2} < 0$ it follows that $T < 0$,

i.e., the machine is a <u>turbine</u>.

From the figure it follows that

$\tan\theta = \frac{U_1}{V_1} = \frac{30\frac{ft}{s}}{20\frac{ft}{s}}$ or $\theta = 56.3°$

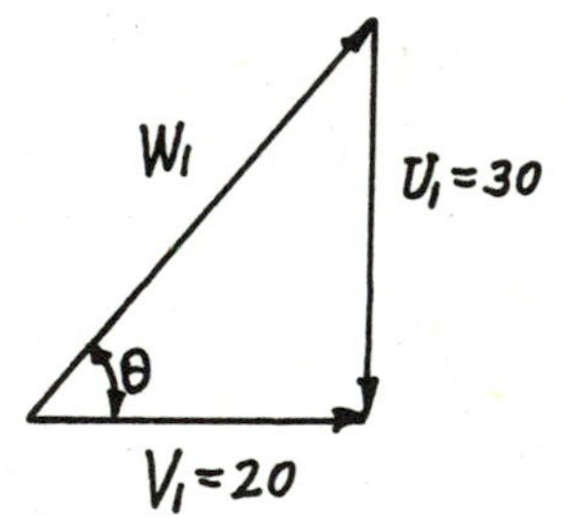

Hence, the blade would be shaped as follows:

(2) 60°

56.3° (1)

and

$w_{shaft} = U_2 V_{\theta 2} = (30\frac{ft}{s})(-1.229\frac{ft}{s}) = -36.9\frac{ft^2}{s^2}$

12.3R

12.3R (Centrifugal pump) Shown in Fig. P12.3R are front and side views of a centrifugal pump rotor or impeller. If the pump delivers 200 liters/s of water and the blade exit angle is 35° from the tangential direction, determine the power requirement associated with flow leaving at the blade angle. The flow entering the rotor blade row is essentially radial as viewed from a stationary frame.

(ANS: 348 kW)

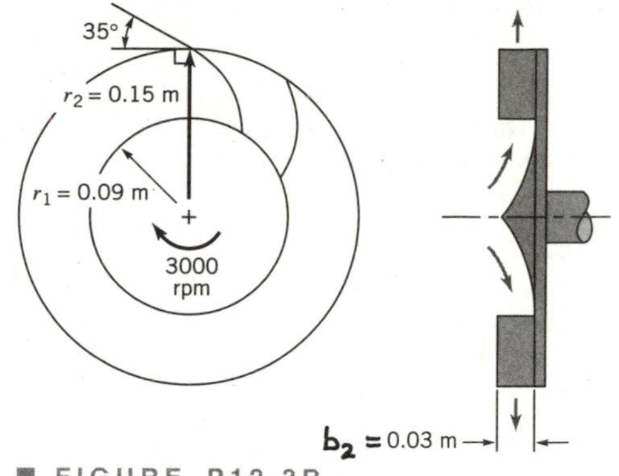

FIGURE P12.3R

From Eq. 12.11 with $V_{\theta 1} = 0$ we have

$$\dot{W}_{shaft} = \rho Q (U_2 V_{\theta 2} - U_1 V_{\theta 1}) = \rho Q U_2 V_{\theta 2} \qquad (1)$$

where

$$U_2 = r_2 \omega = (0.15\,m)\left(3000\,\frac{rev}{min}\right)\left(2\pi\,\frac{rad}{rev}\right)\left(\frac{1}{60}\,\frac{min}{s}\right) = 47.1\,\frac{m}{s}$$

To determine $V_{\theta 2}$ we use the exit velocity triangle shown below.

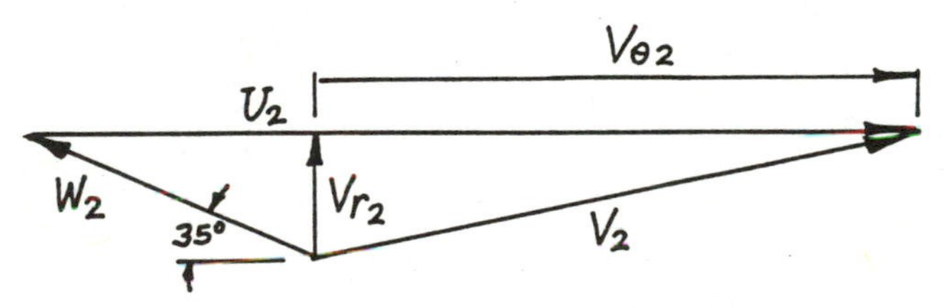

Thus,

$$V_{\theta 2} = U_2 - \frac{V_{r2}}{\tan 35^\circ}$$

with

$$V_{r2} = \frac{Q}{2\pi r_2 b_2} = \frac{\left(200\,\frac{liters}{s}\right)\left(\frac{1\,m^3}{1000\,liters}\right)}{2\pi(0.15\,m)(0.03\,m)} = 7.07\,\frac{m}{s}$$

Hence,

$$V_{\theta 2} = 47.1\,\frac{m}{s} - \frac{7.07}{\tan 35^\circ} = 37.0\,\frac{m}{s}$$

From Eq. (1):

$$\dot{W}_{shaft} = \left(999\,\frac{kg}{m^3}\right)\left(200\,\frac{liters}{s}\right)\left(\frac{1\,m^3}{1000\,liters}\right)\left(47.1\,\frac{m}{s}\right)\left(37.0\,\frac{m}{s}\right)$$

$$= 3.48 \times 10^5\,\frac{N \cdot m}{s} = \underline{\underline{348\,kW}}$$

12.4R

12.4R (Centrifugal pump) The velocity triangles for water flow through a radial pump rotor are as indicated in Fig. P12.4R. **(a)** Determine the energy added to each unit mass (kg) of water as it flows through the rotor. **(b)** Sketch an appropriate blade section.

(ANS: 404 N·m/kg)

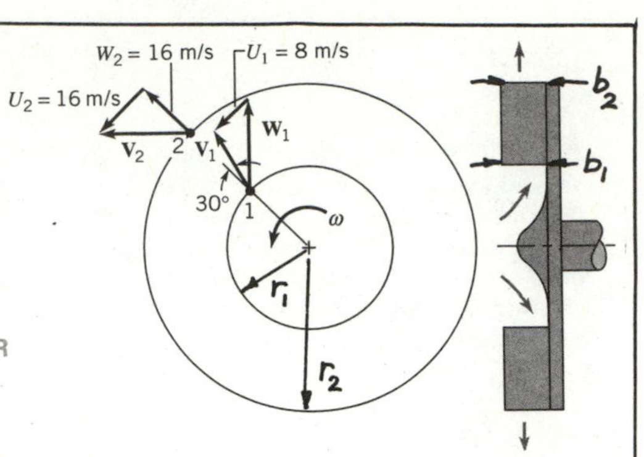

■ FIGURE P12.4R

a) From Eq. 12.12,

$$w_{shaft} = U_2 V_{\theta 2} - U_1 V_{\theta 1}, \quad (1)$$

where since the relative velocity at the exit is in the radial direction (see the figure below), $V_{\theta 2} = U_2 = 16 \frac{m}{s}$

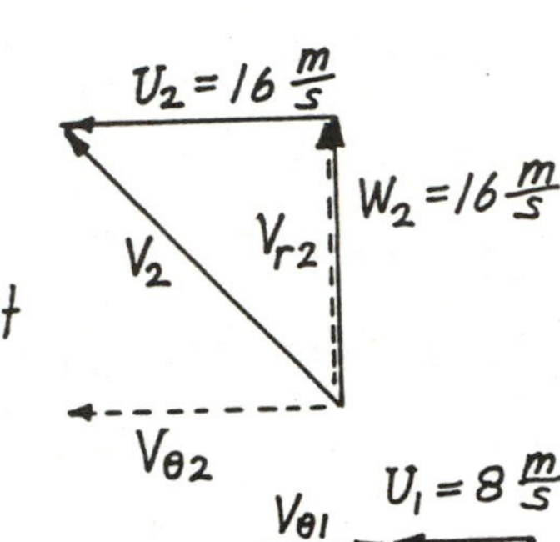

Also, from the inlet conditions,

$V_{\theta 1} = -V_{r1} \tan 30°$, where the minus sign means that $V_{\theta 1}$ is in the opposite direction of U_1

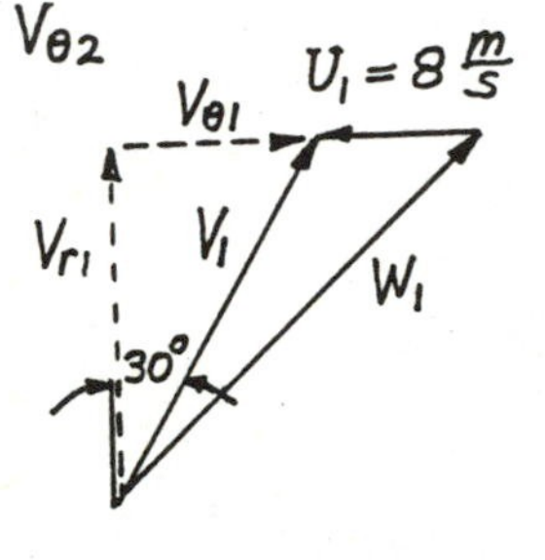

From conservation of mass (with $\rho_1 = \rho_2$) we have

$$V_{r1} A_1 = V_{r2} A_2 = W_2 A_2$$

Thus,

$$V_{r1} = W_2 \frac{A_2}{A_1} = W_2 \frac{2\pi r_2 b_2}{2\pi r_1 b_1} = W_2 \frac{r_2}{r_1} \quad \text{since } b_1 = b_2$$

Also,

$$U_1 = r_1 \omega \text{ and } U_2 = r_2 \omega, \text{ or } \frac{r_1}{r_2} = \frac{U_1}{U_2} = \frac{8 \frac{m}{s}}{16 \frac{m}{s}} = 0.5$$

Thus, $V_{r1} = (16 \frac{m}{s})/(0.5) = 32 \frac{m}{s}$ so that,

$$V_{\theta 1} = -V_{r1} \tan 30° = -(32 \tfrac{m}{s}) \tan 30° = -18.5 \tfrac{m}{s}$$

From Eq. (1):

$$w_{shaft} = (16 \tfrac{m}{s})(16 \tfrac{m}{s}) - (8 \tfrac{m}{s})(-18.5 \tfrac{m}{s}) = 404 \frac{m^2}{s^2} = 404 \frac{N \cdot m}{kg}$$

b) An appropriate blade section would be tangent to W_1 and W_2 at the inlet and exit. Thus, at the exit the blade would be radial. At the inlet,

$$\beta_1 = \tan^{-1}\left[\frac{U_1 - V_{\theta 1}}{V_{r1}}\right] = \tan^{-1}\left[\frac{8 \frac{m}{s} - (-18.5 \frac{m}{s})}{32 \frac{m}{s}}\right] = 39.6°$$

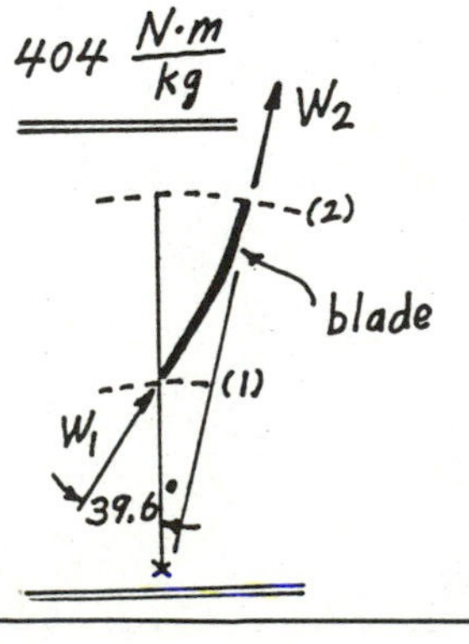

12.5 R

12.5R (Similarity laws) When the shaft horsepower supplied to a certain centrifugal pump is 25 hp, the pump discharges 700 gpm of water while operating at 1800 rpm with a head rise of 90 ft. **(a)** If the pump speed is reduced to 1200 rpm, determine the new head rise and shaft horsepower. Assume the efficiency remains the same. **(b)** What is the specific speed, N_{sd}, for this pump?

(ANS: 40 ft, 7.41 hp: 1630)

a) From Eq. (12.37), for a given pump operating at different speeds

$$\frac{h_{a1}}{h_{a2}} = \frac{\omega_1^2}{\omega_2^2} \quad \text{so that}$$

$$h_{a2} = \left(\frac{\omega_2}{\omega_1}\right)^2 h_{a1} = \left(\frac{1200\,rpm}{1800\,rpm}\right)^2 (90\,ft) = \underline{\underline{40.0\,ft}}$$

Similarly from Eq. (12.38),

$$\frac{\dot{W}_{shaft1}}{\dot{W}_{shaft2}} = \frac{\omega_1^3}{\omega_2^3} \quad \text{so that}$$

$$\dot{W}_{shaft2} = \left(\frac{\omega_2}{\omega_1}\right)^3 \dot{W}_{shaft1} = \left(\frac{1200\,rpm}{1800\,rpm}\right)^3 (25\,hp) = \underline{\underline{7.41\,hp}}$$

b) From Eq. (12.44),

$$N_{sd} = \frac{\omega(rpm)\sqrt{Q\,(gpm)}}{[h_a\,(ft)]^{3/4}}$$

so that

$$N_{sd_1} = \frac{(1800\,rpm)\sqrt{700\,gpm}}{(90\,ft)^{3/4}} = \underline{\underline{1630}}$$

Note that for conditions of state (2), from Eq. (12.36)

$$\frac{Q_1}{Q_2} = \frac{\omega_1}{\omega_2} \quad \text{or} \quad Q_2 = \frac{\omega_2}{\omega_1} Q_1 = \left(\frac{1200\,rpm}{1800\,rpm}\right)(700\,gpm) = 467\,gpm$$

Thus,

$$N_{sd_2} = \frac{(1200\,rpm)\sqrt{467\,gpm}}{(40.0\,ft)^{3/4}} = 1630$$

That is, $N_{sd_1} = N_{sd_2}$. The specific speed is constant for the pump regardless of its speed when operating at a constant efficiency.

12.6R

12.6R (Specific speed) An axial-flow turbine develops 10,000 hp when operating with a head of 40 ft. Determine the rotational speed if the efficiency is 88%.

(ANS: 65.4 rpm)

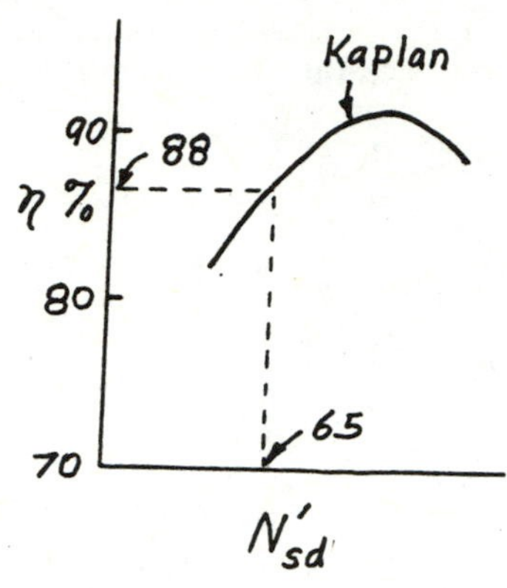

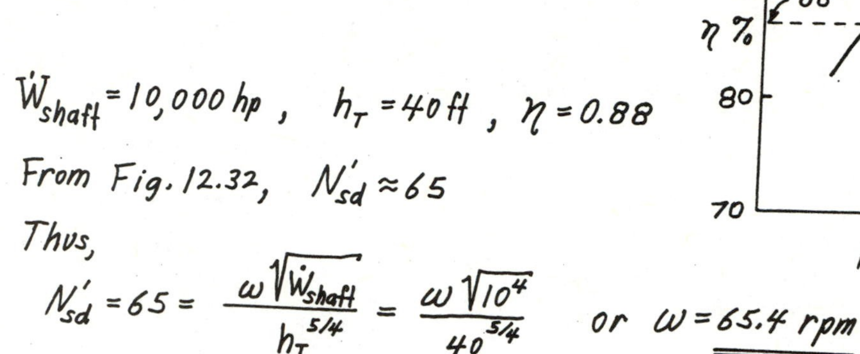

12.7R

12.7R (Turbine) A water turbine with radial flow has the dimensions shown in Fig. P12.7R. The absolute entering velocity is 50 ft/s, and it makes an angle of 30° with the tangent to the rotor. The absolute exit velocity is directed radially inward. The angular speed of the rotor is 120 rpm. Find the power delivered to the shaft of the turbine.

(ANS: −1200 hp)

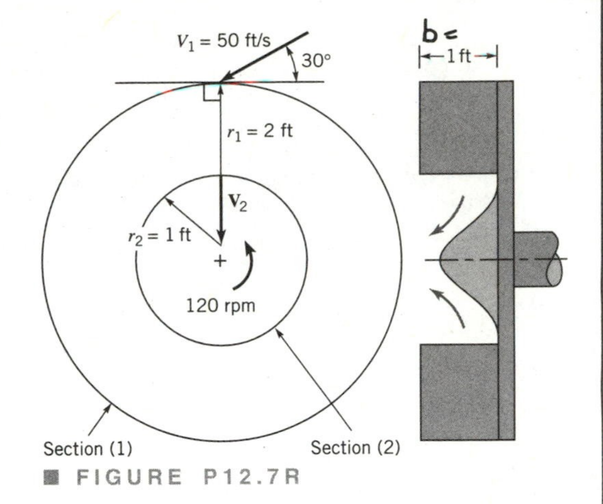

FIGURE P12.7R

$$\dot{W}_{shaft} = \rho Q (U_2 V_{\theta 2} - U_1 V_{\theta 1}) \quad \text{where } V_{\theta 2} = 0 \text{ and} \qquad (1)$$

$$V_{\theta 1} = \left(50 \frac{ft}{s}\right) \cos 30° = 43.3 \frac{ft}{s}$$

Also,

$$Q = 2\pi r_1 b V_1 \sin 30° = 2\pi (2 ft)(1 ft)\left(50 \frac{ft}{s}\right) \sin 30° = 314 \frac{ft^3}{s}$$

and

$$U_1 = \omega r_1 = \left(120 \frac{rev}{min}\right)\left(\frac{1 min}{60 s}\right)\left(\frac{2\pi\, rad}{rev}\right)(2 ft) = 25.1 \frac{ft}{s}$$

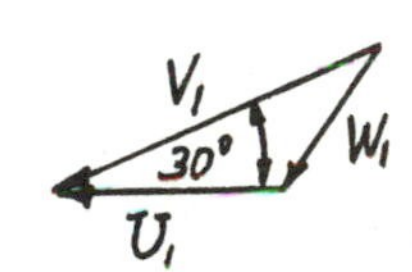

Thus, Eq. (1) gives

$$\dot{W}_{shaft} = \left(1.94 \frac{slugs}{ft^3}\right)\left(314 \frac{ft^3}{s}\right)\left(-25.1 \frac{ft}{s}\right)\left(50 \cos 30° \frac{ft}{s}\right) = -6.62 \times 10^5 \frac{ft \cdot lb}{s}$$

or $$\dot{W}_{shaft} = \left(6.62 \times 10^5 \frac{ft \cdot lb}{s}\right)\left(\frac{1 hp}{550 \frac{ft \cdot lb}{s}}\right) = \underline{\underline{-1,200 \; hp}}$$

12.8R

12.8R (Turbine) Water enters an axial-flow turbine rotor with an absolute velocity tangential component, V_θ, of 30 ft/s. The corresponding blade velocity, U, is 100 ft/s. The water leaves the rotor blade row with no angular momentum. If the stagnation pressure drop across the turbine is 45 psi, determine the efficiency of the turbine.

(ANS: 0.898)

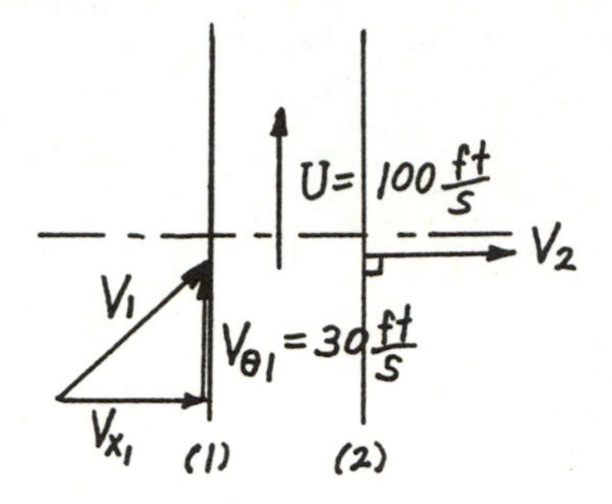

The power removed from the fluid is $\dot{W}_r$, where

$\dot{W}_r = \dot{m} g\left(\frac{\Delta p_s}{\gamma}\right)$, where $\frac{\Delta p_s}{\gamma}$ is the stagnation pressure head drop.

Thus,

$$\dot{W}_r = \rho Q\, g\left(\frac{\Delta p_s}{\rho g}\right) = Q\,\Delta p_s$$

The power removed by the turbine is $\dot{W}_t$, where

$$\dot{W}_t = \rho Q U (V_{\theta 2} - V_{\theta 1}) = -\rho Q U V_{\theta 1} \quad \text{since } V_{\theta 2} = 0$$

Hence,

$$\eta = \text{efficiency} = \frac{|\dot{W}_t|}{\dot{W}_r} = \frac{\rho Q U V_{\theta 1}}{Q\,\Delta p_s} = \frac{\rho U V_{\theta 1}}{\Delta p_s}$$

or

$$\eta = \frac{\left(1.94\,\frac{\text{slugs}}{\text{ft}^3}\right)\left(100\,\frac{\text{ft}}{\text{s}}\right)\left(30\,\frac{\text{ft}}{\text{s}}\right)}{\left(45\,\frac{\text{lb}}{\text{in.}^2}\right)\left(144\,\frac{\text{in.}^2}{\text{ft}^2}\right)} = \underline{\underline{0.898}}$$

TABLE 1.3

Conversion Factors from BG and EE Units to SI Units[a]

	To Convert from	to	Multiply by
Acceleration	ft/s^2	m/s^2	3.048 E − 1
Area	ft^2	m^2	9.290 E − 2
Density	lbm/ft^3	kg/m^3	1.602 E + 1
	$slugs/ft^3$	kg/m^3	5.154 E + 2
Energy	Btu	J	1.055 E + 3
	ft·lb	J	1.356
Force	lb	N	4.448
Length	ft	m	3.048 E − 1
	in.	m	2.540 E − 2
	mile	m	1.609 E + 3
Mass	lbm	kg	4.536 E − 1
	slug	kg	1.459 E + 1
Power	ft·lb/s	W	1.356
	hp	W	7.457 E + 2
Pressure	in. Hg (60 °F)	N/m^2	3.377 E + 3
	lb/ft^2 (psf)	N/m^2	4.788 E + 1
	$lb/in.^2$ (psi)	N/m^2	6.895 E + 3
Specific weight	lb/ft^3	N/m^3	1.571 E + 2
Temperature	°F	°C	$T_C = (5/9)(T_F - 32°)$
	°R	K	5.556 E − 1
Velocity	ft/s	m/s	3.048 E − 1
	mi/hr (mph)	m/s	4.470 E − 1
Viscosity (dynamic)	$lb{\cdot}s/ft^2$	$N{\cdot}s/m^2$	4.788 E + 1
Viscosity (kinematic)	ft^2/s	m^2/s	9.290 E − 2
Volume flowrate	ft^3/s	m^3/s	2.832 E − 2
	gal/min (gpm)	m^3/s	6.309 E − 5

[a]If more than four-place accuracy is desired, refer to Appendix A.